FiWi Access Networks

The evolution of broadband access networks toward bimodal fiber-wireless (FiWi) access networks, described in this book, may be viewed as the endgame of broadband access. After discussing the economic impact of broadband access and current worldwide deployment statistics, all the major legacy wireline and wireless broadband access technologies are reviewed. State-of-the-art GPON and EPON fiber access networks are described, including their migration to next-generation systems such as OCDMA and OFDMA PONs. The latest developments of wireless access networks are covered, including VHT WLAN, Gigabit WiMAX, LTE, and WMN. The advantages of FiWi access networks are demonstrated by applying powerful network coding, heterogeneous optical and wireless protection, hierarchical frame aggregation, hybrid routing, and QoS continuity techniques across the optical–wireless interface. The book is an essential reference for anyone working on optical fiber access networks, wireless access networks, or converged FiWi systems.

Martin Maier is an Associate Professor at the Institut National de la Recherche Scientifique (INRS), University of Québec, and the Founder and Creative Director of the Optical Zeitgeist Laboratory. He received his PhD degree in electrical engineering from the Technical University Berlin, Germany and amongst his awards he was a co-recipient of the 2009 IEEE Communications Society Best Tutorial Paper Award. He is the author of *Optical Switching Networks* (Cambridge University Press, 2008).

Navid Ghazisaidi is an R&D Systems Engineer at Ericsson Inc., San Jose, USA. He received his PhD degree in Telecommunications from the University of Québec, Canada and participated in the prestigious European research projects BIONETS (BIOlogically-inspired autonomic NETworks and Services) and ACCORDANCE (A Converged Copper-Optical-Radio OFDMA-based Access Network with high Capacity and flExibility).

"The area of FiWi networks is central to the current evolution path of networks but presents significant challenges, in particular in integrating disparate systems. This book provides a cogent and highly useful exposition of the main technologies in FiWi, including not only traditional techniques, but also very recent developments such as network coding. This book is a tool both for working engineers and for researchers entering the FiWi area from the optics or from the wireless domains".
Professor Muriel Médard, Massachusetts Institute of Technology

FiWi Access Networks

MARTIN MAIER
Université du Québec, Montréal

NAVID GHAZISAIDI
R&D PDU Broadband Access, Ericsson Inc.

CAMBRIDGE
UNIVERSITY PRESS

University Printing House, Cambridge CB2 8BS, United Kingdom

One Liberty Plaza, 20th Floor, New York, NY 10006, USA

477 Williamstown Road, Port Melbourne, VIC 3207, Australia

314-321, 3rd Floor, Plot 3, Splendor Forum, Jasola District Centre, New Delhi - 110025, India

79 Anson Road, #06-04/06, Singapore 079906

Cambridge University Press is part of the University of Cambridge.

It furthers the University's mission by disseminating knowledge in the pursuit of
education, learning and research at the highest international levels of excellence.

www.cambridge.org
Information on this title: www.cambridge.org/9781107003224

First published 2012

A catalogue record for this publication is available from the British Library

Library of Congress Cataloging in Publication data
Maier, Martin, 1969–
FiWi access networks / Martin Maier, Navid Ghazisaidi.
p. cm.
Includes bibliographical references and index.
ISBN 978-1-107-00322-4 (hardback)
1. FiWi access networks. 2. Wireless communication systems. 3. Optical fiber
communication. I. Title.
TK5105.775
[.M34 2012]
621.39′81–dc23

2011035787

ISBN 978-1-107-00322-4 Hardback

To my parents
M.M.
To the love of my life
N.G.

Contents

Figures

Tables

Tables

Preface

Fiber-wireless (FiWi) access networks may be viewed as the endgame of broadband access. FiWi access networks aim at leveraging on the respective strengths of emerging next-generation optical fiber and wireless access technologies and smartly merging them into future-proof broadband solutions. Currently, many research efforts in industry, academia, and various standardization bodies focus on the design and development of next-generation broadband access networks, ranging from short-term evolutionary next-generation passive optical networks with coexistence requirements with installed fiber infrastructures, so-called NG-PON1, to mid-term revolutionary disruptive optical access network architectures without any coexistence requirements, also known as NG-PON2, all the way to 4G mobile WiMAX and cellular long term evolution (LTE) radio access networks. To deliver peak data rates of up to 200 Mb/s per user and realize what some people refer to as the *vision of complete fixed-mobile convergence* (Ali *et al.* [2010]) it is crucial to replace today's legacy circuit-switched wireline and microwave backhaul technologies with integrated FiWi broadband access networks. To unleash the full potential of FiWi access networks, emerging optical and wireless access network technologies have to be truly integrated at the physical, data link, network, and/or service layers instead of simply mixing and matching them. An interesting example of integrated FiWi access networks is the use of orthogonal frequency division multiplexing (OFDM), which has been successfully deployed in wireless networks but is only recently making its way into PONs, not only to provide a number of desirable characteristics, e.g., increased aggregate bandwidth, scalability, longer reach, lower equipment cost/complexity, and lower power consumption, but also to enable the convergence of a wide range of diverse broadband access technologies, including GPON, EPON, WiMAX, LTE, HFC, and xDSL (Kanonakis *et al.* [2010], Milosavljevic *et al.* [2010]).

This book comprehensively describes the state of the art and latest developments of FiWi access networks from a multitude of perspectives. It starts out with introducing the new definition of the term broadband as outlined by the FCC in its latest broadband deployment report released on July 20, 2010, and then elaborates on the economic impact of broadband access on individuals and enterprises and society at large. Next, we highlight the major findings of OECD's latest report on broadband coverage, taking into account the most important wireline and wireless broadband technologies such as xDSL, cable modem, fiber-to-the-home/building (FTTH/B), broadband over power line (BPL), satellite, etc. After describing current broadband deployments across OECD countries, we summarize recent trends and discuss which broadband access technologies will play an increasingly important role over the next couple of decades and which won't. The second part of the introduction reviews the most important fixed wireline as well as fixed and mobile wireless legacy broadband technologies, including free space optics and UMTS among others, and discusses their pros and cons.

The second and third parts of the book are intended to set the stage and explain the technical details of state-of-the-art fiber and wireless access networks, respectively. More precisely, Part II describes at length both GPON and EPON and introduces the most promising NG-PON1 and NG-PON2 candidates such as XG-PON, long-reach and wavelength division multiplexing (WDM) PON, optical code division multiple access (OCDMA) PON, and OFDMA PON. Part III provides the reader with in-depth information about the latest developments of WiFi, WiMAX, LTE, and wireless mesh networks.

The fourth and final part of the book is dedicated to FiWi access networks. In Part IV, we first elaborate on the difference between conventional radio-over-fiber (RoF) and so-called radio-and-fiber (R&F) networks and their underlying enabling technologies. To learn about their technological maturity and better understand their respective shortcomings, we report on state-of-the-art RoF and R&F testbeds and identify remaining challenges and open issues. We survey previously proposed FiWi access network architectures, ranging from moving cellular network to SuperMAN architectures, and then delve into the technical details of FiWi access networks. We investigate various network planning and reconfiguration techniques to optimize the placement of optical network units (ONUs) and inter-ONU communications. Furthermore, we perform a comparative techno-economic analysis of EPON and WiMAX, which are two key building blocks of FiWi access networks, and investigate their performance for urban, suburban, and rural areas. We look into how and to what extent network coding can be exploited to enhance the performance of NG-PONs and FiWi access networks. Another interesting aspect of bimodal FiWi access networks is their capability to reroute traffic through a wireless mesh front-end in order to improve their survivability. Toward this end, we study the merits and limitations of optical and wireless protection schemes by means of probabilistic analysis for various failure scenarios. In next-generation wireless local area networks (WLANs), frame aggregation is the major performance enhancing mechanism at the medium access control (MAC) sublayer. To further improve the performance of WLAN-based FiWi access networks, we examine novel hierarchical frame aggregation techniques, which are particularly beneficial in carrying video traffic more efficiently. We describe the state of the art of wireless and integrated routing algorithms that aim at optimizing the performance of FiWi access networks in terms of delay, throughput, packet loss, load balancing, and other important metrics such as path availability and power consumption. In addition, we elaborate on various techniques to provide service differentiation and end-to-end QoS continuity across the optical–wireless interface of FiWi access networks. Finally, we would like to point the interested reader to new exciting opportunities of adopting FiWi broadband access networks in other relevant economic sectors such as energy and transportation in order to convert the traditional electric power grid, the largest man-made CO_2 emission source, into the future *smart grid* and thereby enhance the efficiency of energy use and achieve a dramatically increased overall CO_2 reduction across different sectors.

Acknowledgments

This book would not have been possible without the help and contributions of many of our collaborators and colleagues. We would like to thank Dr. Mohammad S. Kiaei for his concise description of the IEEE standard 802.3av covering the salient features of the new physical layer of high-speed 10 Gb/s Ethernet passive optical network (10G-EPON). We are grateful to Professor Chadi M. Assi from Concordia University, Montréal, for his fruitful collaboration in surveying the state of the art of FiWi access network architectures. In particular, we would like to thank Dr. Francesco Paolucci from Scuola Superiore Sant'Anna, Pisa, Italy, for his contributions to the design and performance evaluation of SuperMAN during his four-month research visit at INRS, Montréal. We are especially grateful to Dr. Kerim Fouli and Professor Muriel Médard from the Massachusetts Institute of Technology (MIT), Cambridge, USA, for their excellent work on network coding in next-generation PONs (NG-PONs). We also would like to thank Professor Michael Scheutzow from Technical University of Berlin, Germany, for his probabilistic analysis of the survivability of FiWi access networks and insightful discussions while visiting INRS. At Cambridge University Press, we would like to thank Mia Balashova, Sarah Finlay, and Dr. Phil Meyler for their great support and guidance throughout the whole process of preparing the manuscript. Finally, and most importantly, special thanks go to Martin's beautiful wife Alexie for her patience, love, and belief, and their two wonderful children, who have not missed a single opportunity to play under their dad's desk when he was trying to work on the manuscript at home with more or less success. Navid would like to take this opportunity to express his deep gratitude and appreciation to his parents and his three brothers for their support, love, and encouragement.

Part I

Introduction

1 Broadband access

Access networks connect business and residential subscribers to the central offices of service providers, which in turn are connected to metropolitan area networks (MANs) or wide area networks (WANs). Access networks are commonly referred to as the *last mile* or the *first mile*, whereby the latter term emphasizes their importance to subscribers. Future first-mile solutions have to not only meet the cost sensitivity constraints of access networks arising from the small number of cost-sharing subscribers but also have to provide an ever increasing amount of capacity due to emerging bandwidth-hungry multimedia applications such as video on demand (VoD), high-definition television (HDTV), digital cinema, split-screen video, and 3D online games. These new services and applications are expected to require data rates of up to 100 Mb/s per home, which cannot be provided by traditional narrowband access solutions, e.g., dial-up connections. To meet the bandwidth requirements of emerging and future video-dominated services and applications, legacy access networks have been replaced with broadband access networks over the last few years.

1.1 Definition

The term *broadband* is commonly used to refer to high-speed Internet access with data rates exceeding those of traditional dial-up Internet connections, which typically offer data rates of only 64 kb/s or below. More specifically, the Federal Communications Commission (FCC) used to define broadband service as data transmission speeds exceeding 200 kb/s in at least one direction, i.e., downstream (from the Internet to the subscriber's computer) or upstream (from the user's computer to the Internet). A connection offering at least 200 kb/s in both directions was deemed an *advanced service line* by the FCC. In its latest broadband deployment report released on July 20, 2010, the FCC took the overdue step of raising the minimum speed threshold for broadband in order to respond to evolved broadband technologies and consumer applications and expectations. This change has become necessary due to the fact that most websites feature rich graphics and many embed video. Accordingly, the FCC now defines broadband service as downstream speeds of at least 4 Mb/s and upstream speeds of at least 1 Mb/s. These updated minimum speeds are deemed necessary to stream a high-quality video while leaving sufficient bandwidth for basic web browsing and email (Federal Communications Commission [2010]).

1.2 Economic impact

According to the FCC, broadband enables individuals and enterprises to access a wide range of resources, services, and products related to education, culture, entertainment, telemedicine, e-commerce, public safety, and homeland security. The impact of broadband access networks on the economy was recently investigated in a detailed study carried out by the Organisation for Economic Co-operation and Development (OECD) (OECD [2007]). According to this study, broadband enables the emergence of new business models, processes, inventions as well as improved goods and services. Broadband increases competitiveness and flexibility in the economy by the increased diffusion of information at lower cost and by improving market access to increasingly larger markets, allowing people to work from multiple locations with flexible hours and speeding up procedures and processes. In fact, broadband may be viewed as a so-called *general purpose technology (GPT)* that has the potential to fundamentally change how and where economic activity is organized.

As discussed in greater detail in (OECD [2007]), the impact of providing residential and business subscribers with broadband access is manifold. Among others, broadband in conjunction with other advanced information and communication technologies (ICTs) facilitates the globalization of services, e.g., legal, accounting, advertising, design, software programming, IT support, management consultancy, human resource management, and labour recruitment services. The broadband and ICT enabled globalization of services has a fundamental impact on the way economies work as firms focus on their core competitive advantage activities and outsource/offshore the rest. However, the overall effect of broadband on employment is ambiguous. While broadband is thought to enhance growth and thus create employment opportunities, it also facilitates capital–labour substitution. On the other hand, with broadband access, people outside the boundaries of traditional institutions and hierarchies can also innovate to produce new goods and services. Broadband connections also increasingly allow people to carry out tasks at times and places they could previously not, facilitated by a corporate culture that is increasingly output-oriented rather than location- or time-oriented. While increased flexibility is likely to have a positive impact on workers' productivity, companies can also save costs on office space. Telework is a significant phenomenon and with increasing broadband penetration it can be expected to grow further. Telework is often thought to improve flexibility and quality of life for workers, relieve traffic congestion problems, and reduce pollution. With broadband access in their homes, teleworkers are able to work away from the office with alternative working patterns better suited to the demands of their lifestyle and commitments than the traditional 9–5 office hours, resulting in a better work–life balance. Flexible working practices enabled by broadband access might help increase labour market participation and reduce problems related to transport, environment, and aging populations, as well as help ease bottlenecks in the health sector, e.g., monitoring patients in rural areas. In addition, broadband not only changes the role of individuals in the productive process, facilitating user-created content (UCC), also known as user-generated content (UGC), and user-driven innovation, but also has a positive impact on consumer surplus. Broadband access networks bring about lower search and information cost and greater access to information, which makes price comparisons easier, increases competition, and creates a downward pressure on prices. They also enable the increased customization of goods and services and the ensuing improved quality. This process not only affects online shopping but also can affect offline shopping when people search for information and compare prices online before going to a shop to make the actual purchase. The resulting empowerment of the individual through the use of broadband access networks constitutes a very important economic impact and brings about a restructuring of the relationship between customers and suppliers. There are new broadband-enabled services that give consumers direct access to lower cost services abroad. An example given in (OECD [2007]) is an Indian company that offers tutoring services to pupils for less than half the price of those available in the United Kingdom. All that is needed is a broadband connection, a headset, and the software provided by the company. Broadband access also helps empower small and medium enterprises (SMEs), enabling them to compete with larger firms in an increasing number of markets and to purchase services they previously could not afford, e.g., remote IT services such as website hosting, voice over Internet protocol (VoIP) service, network security, or software as a service (SaaS) rather than hiring an IT technician. As a result, broadband access increasingly enables people to start small businesses from home, which will contribute to a more dynamic and entrepreneurial business sector.

1.3 Coverage

In December 2009, the OECD published its latest report on broadband coverage across OECD countries taking the most important fixed wired and wireless broadband technologies into consideration (OECD [2009]). The reported coverage data indicate the extent to which businesses and residential customers have access to broadband, i.e., to what extent the population and businesses are able to subscribe to broadband if they wish. Historically, for the purpose of data collection, the OECD has considered broadband as a service providing Internet access at speeds higher than 256 kb/s. The broadband technologies considered in this report include all access technologies except mobile broadband, whose measurement methodology is currently being developed. Specifically, the report includes the following fixed wireline and wireless broadband technologies (to be described in technically greater detail in Chapter 2):

Fixed wireline broadband technologies

- Full variety of digital subscriber line (DSL)
- Cable modem
- Fiber-to-the-premises (FTTP), i.e., fiber-to-the-home and fiber-to-the-building (FTTH/B)
- Broadband over power line (BPL)

Fixed wireless broadband technologies

- Satellite
- Worldwide interoperability for microwave access (WiMAX)

In addition, the report contains data on 3G coverage focusing on the two mobile wireless technologies listed below:

3G mobile wireless technologies

- Wideband code division multiple access (WCDMA) for universal mobile telecommunications system (UMTS)
- CDMA2000

In the following, we summarize the major findings of OECD's latest report on broadband coverage for the aforementioned fixed wireline/wireless broadband and 3G mobile technologies. In our discussion, we also include some illustrative broadband-related statistics from the OECD broadband portal (OECD [2010]). Figure 1.1 depicts the total number of broadband subscribers (in million) for various OECD countries. Population distribution patterns, both in terms of density and dispersion, and difficult terrain are among the most important factors affecting broadband coverage. Low population density and high dispersion can encourage the use of terrestrial wireless or satellite technology (and/or a hybrid of these) rather than wired infrastructure that may be more expensive and less timely to deploy. Figure 1.2 shows the broadband subscriptions sorted by technology for a total of 271 million subscribers. DSL and its variants, in particular asymmetric DSL (ADSL), account for the majority (60%) of broadband subscriptions in OECD countries. Cable modem Internet access is the broadband technology with the second largest number of subscribers across OECD countries, representing 29% of the OECD broadband subscription base. Fiber connections are still a minority in almost every OECD country, accounting for only 9% of

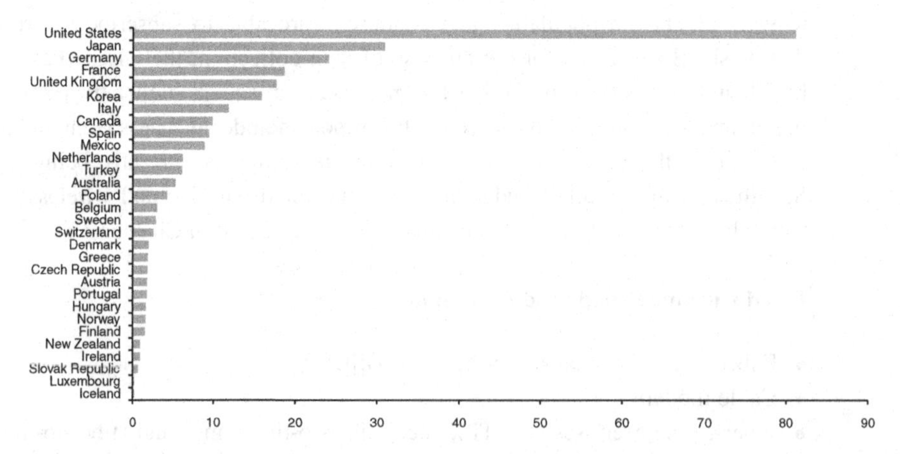

Figure 1.1 Total fixed and wireless broadband subscribers by country, OECD Broadband Portal, http://www.oecd.org/sti/ict/broadband, accessed on 16/06/11.

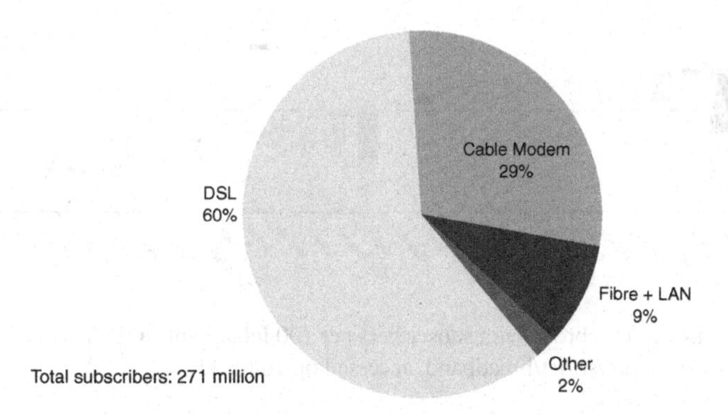

Figure 1.2 Fixed and wireless broadband subscriptions by technology, OECD Broadband Portal, http://www.oecd.org/sti/ict/broadband, accessed on 16/06/11.

current broadband subscriptions (under FTTB, local area network [LAN] deployment in apartments is included). However, it is important to note that an increasing number of telecommunications operators are starting to launch this service to the market and coverage is expected to increase in the short to medium term. This will represent an important step towards higher transmission speeds and the opportunity for new services, which may bring into question the economic viability of some of the existing broadband technologies. Only 2% of current broadband subscribers use technologies other than DSL, cable modem, and FTTH/B. Satellite broadband solutions using the Ka band are starting to be deployed by providers and the future is promising in terms of performance as bit rates between 10 and 20 Mb/s are expected. Ka band technology reduces the required dish size, resulting in a lower equipment cost to consumers. It employs so-called spot beams rather than regional or hemispheric transponder coverage, allowing for a more efficient use of bandwidth. However, in the long run, the performance of Ka band satellite systems will be lower than that of wired broadband technologies, not only in terms of bit rate but also latency and price. Satellite broadband requires expensive customer premises equipment (CPE), including a satellite dish, and it is more expensive than other broadband access technologies. Moreover, it suffers from a high latency and its performance may be seriously affected by weather conditions. It is clear that satellite suppliers will hardly reach the price to speed ratio that wired broadband suppliers offer unless they are subsidized. This is actually the case in some rural broadband deployment plans, where satellite connections are offered at comparable prices to those of wired broadband. For instance, Australian satellite operators offer coverage to almost 100% of the Australian population. This is a solution very well adapted to a vast continent like Australia, where terrestrial technology cannot provide broadband services to all the population, especially in remote areas. WiMAX has been regarded as a key technology to provide broadband services in rural areas. Satellite and fixed wireless broadband technologies are often potentially good solutions for improved coverage since their deployment costs are far less than those of wired solutions. However, while satellite technologies mostly reach 100% population coverage, WiMAX needs base station deployments, which may hinder its success in rural areas

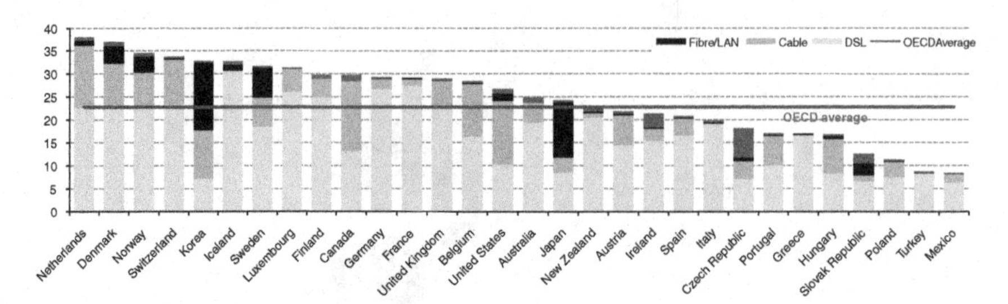

Figure 1.3 Fixed and wireless broadband subscribers per 100 inhabitants, OECD Broadband Portal, http://www.oecd.org/sti/ict/broadband, accessed on 16/06/11.

not reached by wired broadband technologies. Lastly, BPL has not yet been used for delivering extended broadband coverage, even though the ubiquity of power networks would in principle suggest its use as an important technology in rural areas. However, BPL faces a number of technological and regulatory challenges that have hindered its development. Currently, BPL accounts for less than 30 000 subscribers across OECD countries.

Figure 1.3 provides more detailed information about the numbers of broadband subscribers per 100 inhabitants for the three predominant broadband technologies (DSL, cable modem, and FTTH/FTTB) in OECD countries with the highest broadband penetration. DSL is the most common broadband technology used for high-speed Internet connections in most OECD countries. Cable networks were initially deployed in OECD countries to support cable television service, but have been extensively upgraded in order to provide cable modem broadband access to the Internet. Cable modem coverage varies significantly across OECD countries. Italy and Greece have no cable modem as historically they do not have cable television. Coverage in Turkey and New Zealand is also very limited. In contrast, the United States, Canada, Belgium, the Netherlands, and Portugal enjoy quite a high cable modem coverage. In fact, Canada and the United States are the only two countries where cable modem is leading in terms of broadband coverage, surpassing even DSL coverage. Cable broadband providers are starting to upgrade their network infrastructure to adopt data over cable service interface specifications (DOCSIS) 3.0, the new version of DOCSIS that will enable download speeds of around 150 Mb/s, which may well provide significant competition to fiber deployment in areas where there is infrastructure competition. The main concern about the future of cable networks is how they will face competition from fiber infrastructure and whether DOCSIS 3.0 standard developments will be able to face FTTH/B competition, which in principle can reach higher transmission rates. However, DOCSIS 3.0 upgrade is less resource demanding than fiber deployment in terms of cost since it relies on an existing infrastructure. Note that Fig. 1.3 also shows the OECD average number of DSL, cable modem, and FTTH/B subscribers, which is equal to 22.8 per 100 inhabitants.

Figure 1.4 depicts the percentage of fiber connections (FTTH/B) in total broadband subscriptions, while FTTH/B household coverage is shown in Fig. 1.5. In particular, Japan and (South) Korea have high coverage of fiber broadband. About 86.5% of Japanese and 67% of Korean households have FTTH/B coverage. Sweden, France,

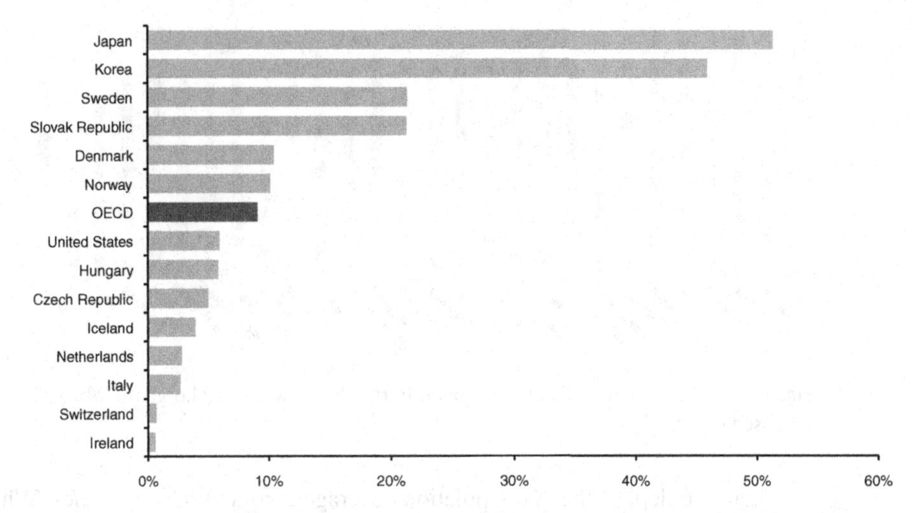

Figure 1.4 Percentage of fiber connections in total broadband, OECD Broadband Portal, http://www.oecd.org/sti/ict/broadband, accessed on 16/06/11.

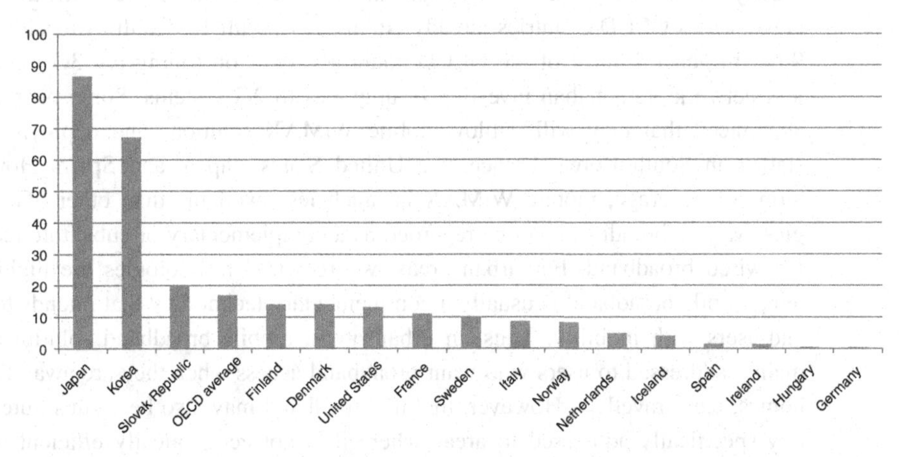

Figure 1.5 FTTH/B coverage, OECD Broadband Portal, http://www.oecd.org/sti/ict/broadband, accessed on 16/06/11.

the United States, Denmark, and Finland reach an FTTH/B household coverage above 10%. In the Slovak Republic, FTTH/B coverage is 20%. On the other hand, 14 OECD countries have no FTTH/B coverage and four other OECD countries have less than 2% of households passed. However, this situation might rapidly change if some of the largest telecommunications operators start to launch their fiber offers on a broader scale, as Verizon in the United States, Numericable in France, or Fastweb in Italy have already done. As a matter of fact, fiber is starting to gain importance in OECD countries. Although FTTH/B household coverage is still low for most of them with the OECD average currently being equal to 16.96%, operators are expected to increase their fiber rollouts in the short to medium term.

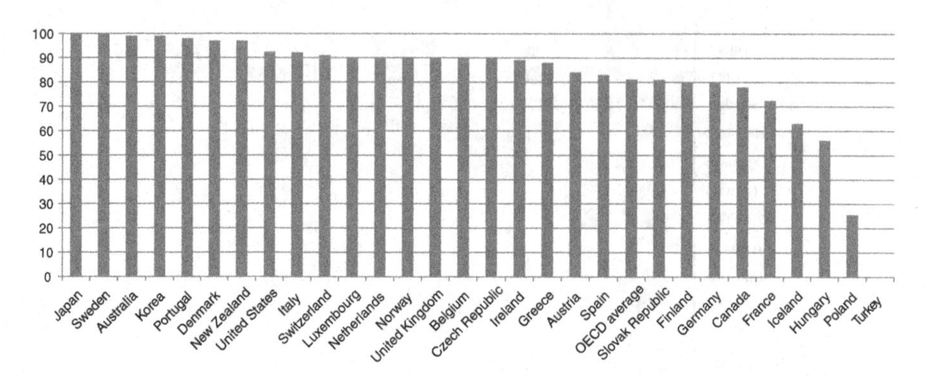

Figure 1.6 3G coverage, OECD Broadband Portal, http://www.oecd.org/sti/ict/broadband, accessed on 16/06/11.

Figure 1.6 depicts the 3G population coverage across OECD countries. While some countries have almost achieved complete coverage, e.g., Japan reaching 100% coverage already in March 2007, Sweden (100%), Australia (99%), and South Korea (99%), Turkey has not started its 3G deployment yet. The population-weighted average coverage across OECD countries already attains 81%, with 16 countries reaching at least 90%. In general, most of the OECD countries focus on rolling out 3G mobile wireless networks rather than investing in upgrades to 2G systems. Some operators have announced that they will deploy mobile WiMAX solutions and deployments have started in South Korea, France, the United States, Japan, and Spain. However, in terms of coverage, mobile WiMAX is much less available than other 3G technologies. Mobile broadband can be regarded as a complementary or substitute technology for wired broadband. For urban areas, where wired technologies are highly prevalent, mobile broadband is usually a complementary technology that intends to provide end-users with mobility. Thus, in urban areas, mobile broadband solutions may be mainly addressed to users who want broadband access when they are away from their homes, e.g., travellers. However, mobile broadband may also be a substitute technology specifically addressed to areas where it is not economically efficient to deploy wired broadband technologies, such as rural, remote, or scarcely populated areas. For these situations, mobile broadband (together with fixed wireless) technologies may address specific broadband access problems and act as a substitute for wired broadband technologies.

1.4 Forecast

As we have seen in the previous section, the average FTTH/B household coverage in OECD countries is close to 17% at present, while the majority of subscribers resort to DSL or cable modem technologies for broadband access. However, it is important to note that both DSL and cable networks rely on broadband solutions that push fiber deep into the access network, leading to so-called *deep fiber access*

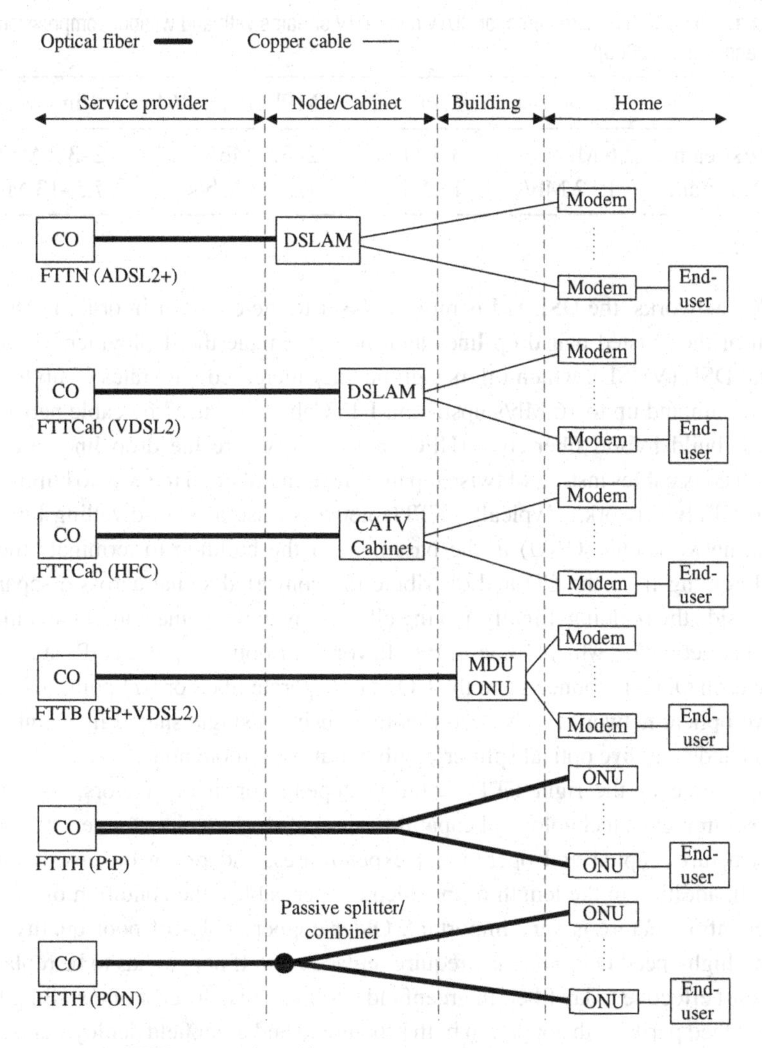

Figure 1.7 FTTx network architectures.

solutions (Kautz and Walker [2005]). We are currently witnessing a strong worldwide deployment of deep fiber access solutions to push optical fiber closer to individual homes and businesses and to help realize FTTx networks, where x denotes the discontinuity between optical fiber and some other transmission medium (Koonen [2006]). Figure 1.7 shows a variety of possible FTTx network architectures, ranging from fiber-to-the-node (FTTN) to FTTH networks, depending on how deep optical fiber reaches into the first/last mile. In FTTN networks, fiber is used between the central office (CO) and the DSL access multiplexer (DSLAM) located at the remote node. The DSLAM connects to each end-user's modem through a separate legacy telephone twisted pair of copper wires. FTTN networks deploy asymmetric DSL2+ (ADSL2+) on the twisted pairs offering data rates of up to 24 Mb/s downstream and up to 3.5 Mb/s upstream. In fiber-to-the-cabinet (FTTCab) networks, also known as fiber-to-the-curb

Table 1.1. Bandwidth requirements for SDTV and HDTV streams with and without compression (Kautz and Walker [2005]).

	Broadcast	MPEG-2	MPEG-4/H.264	Windows Media 9
SDTV stream	6 Mb/s	3.5 Mb/s	2–3.2 Mb/s	2–3.2 Mb/s
HDTV stream	19.2 Mb/s	15 Mb/s	7.5–13 Mb/s	7.5–13 Mb/s

(FTTC) networks, the DSLAM is moved closer to the end-user in order to shorten the length of the twisted-pair drop lines and thereby enable the deployment of very high bit-rate DSL (VDSL), which offers significantly increased data rates of up to 52 Mb/s downstream and up to 16 Mb/s upstream. FTTCab is also used by cable network operators to build hybrid fiber-coax (HFC) networks, where the drop lines are realized through coax cables instead of twisted pairs. Bringing fiber all the way to buildings gives rise to FTTB networks. Typically, FTTB networks use a multi-dwelling unit (MDU) optical network unit (ONU) in the basement of the building to terminate the optical signal coming from the CO and distribute the converted signal across a separate network inside the building. Finally, paving all the way to the home with optical fiber leads to FTTH networks, which come in two flavors: (i) point-to-point (PtP) star topology, where each ONU is connected to the CO via a separate fiber, or (ii) point-to-multipoint passive optical network (PON) tree topology, using a single shared fiber link between the CO and a passive optical splitter/combiner at the remote node.

The choice of the right FTTx solution depends on many factors, including service requirements, technological capabilities, regulatory requirement, existing installed infrastructure, capital and operational expenditures, and return on investment (ROI) goals. In addition to the length of installed copper cables, the condition of the existent copper infrastructure is very important. Older copper cables of poor quality may not support high-speed DSL and may require replacement. If copper has to be replaced, it is more cost effective to lay fiber. In greenfield deployments, the cost of installing fiber has now reached parity with copper. In both brownfield and greenfield deployments, it is less expensive to use fiber instead of copper, given the longer life span of fiber and the corresponding lower ongoing maintenance costs. According to (Wagner *et al.* [2006]), the plant of PON-based FTTH/B networks enables major operational-cost savings, which nearly pay for the installation expenses over a seven-year period without even taking service-related revenues into account. Also with respect to expected bandwidth requirements per home, a deep fiber access solution like FTTH/B will be the best alternative for carrying triple-play traffic (i.e., voice, video, and data) in a future-proof manner. Table 1.1 lists the bandwidth requirements to carry a single standard definition television (SDTV) or HDTV stream with and without using various compression algorithms. The bandwidth requirement for multiple HDTV VoD streams ranges from 40 Mb/s for support of four HDTV channels to 80 Mb/s for support of up to eight HDTV channels per home. If an additional 20 Mb/s for high-speed data and VoIP is allocated, then the total bandwidth required for triple-play services is between 60 Mb/s and 100 Mb/s per home (Kautz and Walker [2005]).

Deep fiber access is a challenging mix of technology choices, business models, and regulatory issues. But due to increasingly personalized, interactive, on-demand, and high-definition broadband service demands and thanks to innovations within low-cost fiber-laying and fiber-blowing techniques, e.g., micro-trenching, there is no doubt that widespread FTTH/B deployment is just a matter of time (Ericsson [2008]).

According to the International Telecommunication Union (ITU), the total number of fixed broadband subscribers worldwide has grown more than threefold over the past 5 years, from about 150 million in 2004 to almost 500 million by the end of 2009. In 2008, China overtook the United States as the largest fixed broadband market in the world. At the end of 2008, China's fixed broadband penetration was 6.2 subscribers per 100 inhabitants, the highest of any low or lower-middle-income economy in Asia and the Pacific. In contrast, in Africa there is only one fixed broadband subscriber for every 1000 people, while in Europe there are 200 subscribers for 1000 people. Clearly, these figures indicate that there exists a serious global *broadband divide* between developed and developing countries. In most developing countries, the price for fixed broadband access remains prohibitively high, preventing people from having access to the information society. Similar observations were made for mobile broadband subscriptions (ITU [2009]).

The impact of income distribution on broadband penetration in developed and developing countries was investigated in (Handler and Grossman [2009]). Broadband penetration rates have soared in most developed countries over the past several years, benefiting gross domestic product (GDP) growth. Broadband affordability is a major inhibitor to growth in developing countries. Beside the per-capita GDP, a country's income distribution can also inhibit growth since in countries with highly skewed income distribution only a limited share of the population can afford broadband. While in developed countries broadband affordability is not a material issue for the vast majority of households, developing countries should consider social policies that promote a more egalitarian income distribution in order to maximize the benefits of future broadband investments. Such policies might include tax breaks for the lowest income classes, particularly for broadband equipment purchases.

In its latest home broadband adoption report, the Pew Internet & American Life Project identifies the major barriers to broadband adoption in the United States. Generally, the two most important factors positively correlated with home broadband adoption are income and education. However, while population subgroups that have above average broadband usage rates saw modest increases from 2008 to 2009, the greatest growth in broadband adoption has taken place in the same time period among low-income Americans and senior citizens, two population subgroups that have below average broadband usage rates. Specifically, Americans age 65 and over had a broadband adoption growth of 58% from 19% to 30%, while Americans living in households with annual incomes of US$ 20 000 or less saw a broadband adoption growth from 25% in 2008 to 35% in 2009, which translates into a 40% growth. Broadband adoption in the United States appears to have been largely immune to the effects of the current economic recession. According to the report, an increasing number of respondents said that they had cut back or canceled a cell phone plan or cable television service rather than their

Internet service. The rise in home broadband adoption, in the face of a severe economic recession, may seem surprising. On the other hand, the migration to the Internet of many resources for finding and applying for jobs may prompt some to cut something else and keep (or add) broadband. It appears that few people are willing to cut back on broadband and people are more likely to economize on communication services other than the Internet. The report also shows that home wireless networking has been steadily on the rise, reaching 37% of all Internet users in 2009 (Horrigan [2009]).

While FTTH currently accounts for only 5% of US subscribers with broadband at home, the take-rate for FTTH services continues to increase. In a recent FTTH North American market update by RVA Market Research & Consulting, take-rates have been fairly steady at 52%. These rates include some cases where fiber has been replaced for all customers, even voice only customers. However, it is important to note that even voluntary take-rates for some individual FTTH projects exceed 82%. This is especially true in rural areas that were previously underserved with both Internet and video services and therefore have little real competition (RVA Market Research & Consulting [2009]).

According to recent market data by ABI Research, the number of fixed broadband subscribers will rise to 501 million by the end of 2014, of which 106 million will subscribe to services delivered via fiber. Among the three major fixed broadband technologies, the number of fiber subscribers will increase fastest at a compound annual growth rate of 20% from 2008 to 2014. Furthermore, FCC has recently unveiled the United States' first national broadband plan with the goal to provide every American with broadband access speeds of 100 Mb/s by 2020. Toward this goal, fiber (together with next-generation wireless) broadband technologies will play an increasingly vital role in future broadband access networks. This is already witnessed by installed state-of-the-art VDSL equipment that is almost exclusively based on optical fiber backhaul solutions. While copper will certainly continue to play an important role in current and near-term broadband access networks, it is expected that FTTH deployment volume will keep increasing gradually and will eventually become the predominant fixed wireline broadband technology by 2035 (Ödling *et al.* [2009]).

2 Legacy broadband technologies

Broadband access involves various enabling technologies. The choice of broadband technology generally depends on a number of factors such as availability, price, location, service bundling, and technological requirements. In this chapter, we review the technical details of the most common types of fixed wireline, fixed wireless, and mobile wireless legacy broadband technologies, including, but not limited to, digital subscriber line, cable modem, and 3G systems. In our discussion, we try to highlight the benefits and limitations of available legacy broadband technologies.

2.1 Fixed wireline broadband technologies

2.1.1 Digital subscriber line

The local subscriber loop of most of today's telephone companies consists of unshielded twisted pair (UTP) copper wires. The length of the local loop depends on a number of factors such as population density and location of the connected residential or business customers. However, it is usually no longer than 4–6 km due to limitations stemming from legacy narrowband telephony. The copper wire pairs can be buried or aerial and are typically grouped together in so-called binders, also known as bundles, which may contain tens or even hundreds or thousands of twisted pairs (Czajkowski [1999]).

Traditional voiceband modems operate at the bottom frequencies (0–4 kHz) of the spectrum available on twisted pairs and offer data rates of no more than 56 kb/s. It is important to note, however, that it is not the UTP copper wires that prevent transport of broadband data signals but rather the bandwidth allocated by legacy telephone company switches to voice calls. Much higher data rates can be achieved by avoiding the narrowband voice switch and instead have a transmission path that only includes the twisted pair. As shown in Fig. 2.1, the family of different digital subscriber line (xDSL) modem technologies makes use of a splitter at both sides of each twisted pair to separate data signals from the plain old telephone service (POTS). At the central office (CO), the extracted data signals are routed as broader bandwidth digital signals through an appropriate broadband network, which may be based on technologies such as asynchronous transfer mode (ATM) and Internet protocol (IP). The data rates achievable over twisted pair phone lines depend heavily on the length of these lines. In general, the capacity of twisted copper wires decreases for increasing length. The currently fastest

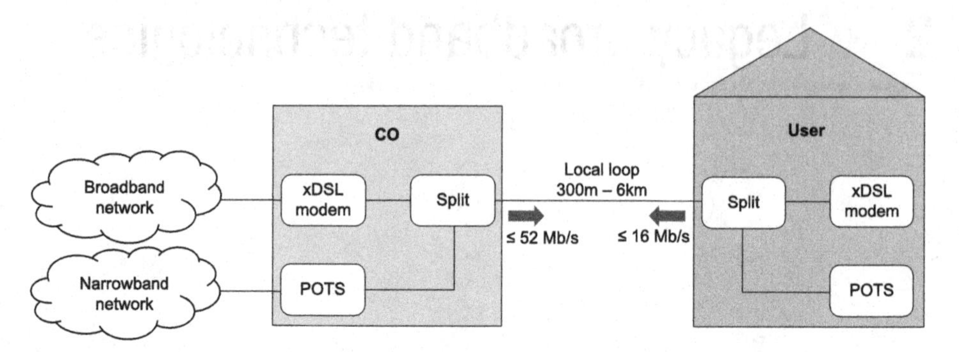

Figure 2.1 Digital subscriber line (xDSL) communication model. After Cioffi *et al.* (1999). ©1999 IEEE.

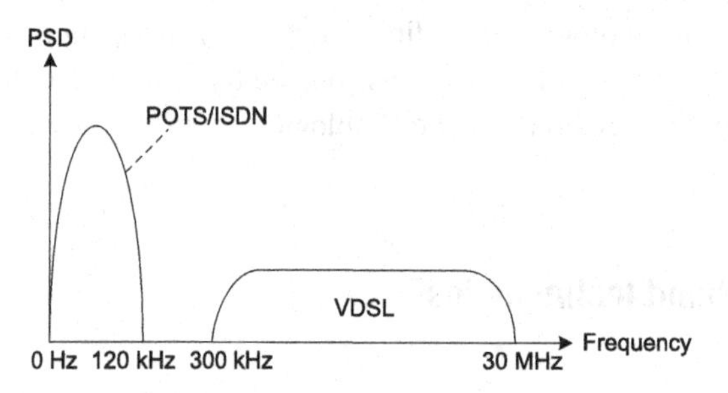

Figure 2.2 Spectrum allocation of VDSL and traditional narrowband POTS/ISDN services. After Cioffi *et al.* (1999). ©1999 IEEE.

DSL technology, also known as very high bit-rate DSL (VDSL), is able to deliver data rates of up to 52 Mb/s downstream and up to 16 Mb/s upstream on the shortest twisted pairs with a length of 300 m or less (Cioffi *et al.* [1999]).

The way xDSL technologies provide increased data rates is by using frequencies above 4 kHz, which have not been used previously for voice transmission due to unacceptable levels of crosstalk. By capitalizing on highly sophisticated techniques enabled by the continuing advancement of lower cost and more powerful digital signal processing (DSP) chips, it became possible to limit the detrimental impact of crosstalk and other physical transmission impairments (e.g., attenuation, radio frequency ingress, thermal noise, and impulsive noise) and therefore expand the bandwidth potential over a single pair of copper wires. These techniques not only permit POTS to continue unaffected over twisted pairs in the same binder, but also permit POTS to continue simultaneously on the same phone line along with xDSL transmission (Humphrey and Freeman [1997]). For illustration, Fig. 2.2 depicts the power spectral density (PSD) of VDSL and traditional narrowband POTS and integrated services digital network (ISDN) services sent across the same phone line, whereby VDSL is able to achieve significantly higher data rates than traditional voiceband modems and ISDN by exploiting frequencies in the range of 300 kHz to 30 MHz well above the frequencies used by narrowband

POTS/ISDN services. To separate the narrowband and broadband signals at both ends of the phone line, each VDSL modem's splitter contains a low-pass filter passing the POTS/ISDN and a high-pass filter passing the VDSL signals.

Depending on the bandwidth requirements of customers, different optical fiber based architectures are possible for xDSL deployment, as we have already seen in Fig. 1.7 when we were discussing the various types of fiber-to-the-x (FTTx) architecture. In fiber-to-the-node (FTTN) and fiber-to-the-cabinet (FTTCab) based xDSL architectures, the DSL access multiplexer (DSLAM) is placed at the remote node and street cabinet, respectively. The DSLAM multiplexes the upstream and downstream transmissions of multiple subscribers, each connected via a separate pair of copper wires. In general, xDSL deployments exhibit a fundamental trade-off between the length of the twisted pair copper wires and the achievable data rate. The nearer the optical fiber reaches to the subscriber, the shorter the length of the copper drop lines and the higher the achievable data rate. According to (Maxwell [1996]), the following downstream data rates can be realized for the indicated local loop lengths:

1.5 Mb/s	18 000 ft
2.0 Mb/s	16 000 ft
6.0 Mb/s	12 000 ft
9.0 Mb/s	9 000 ft
13.0 Mb/s	4 500 ft
26.0 Mb/s	3 000 ft
52.0 Mb/s	1 000 ft

The family of xDSL technologies has been growing over the last decades with the goal to provide ever increasing data rates to business and residential customers, at the expense of requiring progressively more bandwidth in the twisted pair copper wires. The most important xDSL technologies include high-bit-rate DSL (HDSL), asymmetric DSL (ADSL), and VDSL (Humphrey and Freeman [1997], Czajkowski [1999]):

- **HDSL**
  HDSL is the oldest of the xDSL technologies and has been widely deployed since 1993. Originally, HDSL used three pairs of copper wires, but later standards specified a 2-pair HDSL, followed by a version of HDSL using a single copper pair. Most HDSL implementations provide either 1.544 or 2.048 Mb/s of symmetrical bandwidth up to 12 000 ft from the CO. These speeds conform to T1 and E1 standards in North America and Europe, respectively. Hence, HDSL has been primarily deployed for the provisioning of T1/E1 leased lines among business customers.

  The later versions of HDSL use trellis coded pulse amplitude modulation (TC-PAM). HDSL has been attractive because it greatly reduces the cost of traditional T1/E1 lines by eliminating the need for repeaters, as opposed to native T1/E1 lines that require repeaters every 1 km. Furthermore, HDSL is tolerant to poor quality cables and may also be used for rapid deployment.

- **ADSL**

  ADSL allocates more bandwidth for downstream than upstream transmissions. The original version of ADSL was standardized in 1995 and was capable of delivering up to 8 Mb/s downstream and 1 Mb/s upstream. The latest version of ADSL, referred to as ADSL2+, increases the available downstream data rate to 24 Mb/s.

  The majority of commercially available ADSL modems deploy the standard-based *discrete multitone (DMT)* modulation technique. DMT is a multicarrier modulation technique, which divides the phone line into many small channels and modulates each channel for data transmission. An earlier modulation technique widely used in commercially available ADSL modems was the so-called carrierless amplitude phase (CAP) modulation, a technique closely related to quadrature amplitude modulation (QAM). CAP is a single-carrier modulation technique that uses the bandwidth above 4 kHz as a single data transmission channel.

  Channel allocation in ADSL may be done in two modes: (*i*) frequency division multiplexing (FDM) mode, or (*ii*) echo cancellation (EC) mode. Each mode blocks off the lower 25 kHz for POTS (POTS only needs 4 kHz, but splitters are easier to design for higher cut-off frequencies). In FDM mode, the downstream channel starts above the upstream channel at approximately 240 kHz and extends as far up as needed, whereby the upstream channel uses bandwidth in the order of 135 kHz between the downstream and POTS channels. In EC mode, the downstream channel overlaps the upstream channel. This has the advantage that the downstream channel has more bandwidth in the good transmission window of the phone line and the upstream channel can be extended upward, at the expense of an echo canceller (Maxwell [1996]).

  ADSL speeds may vary greatly based on a number of conditions. In areas with a large variance in the length of the local loop, the type of copper wire, and the quality of the phone line, it becomes difficult to determine what speeds should be provisioned over each phone line. *Rate-adaptive DSL (RADSL)* is a subset of ADSL that automatically adjusts the data rate based on a series of initial tests that determine the maximum speed possible on a given phone line, thereby taking much of the effort and/or guesswork out of provisioning ADSL (Humphrey and Freeman [1997]).

- **VDSL**

  VDSL is the most recent and highest-speed xDSL technology, originally standardized in 2001. VDSL is defined to offer asymmetric and symmetric services. The first type of service is usually for residential customers, while the latter one is generally viewed as a business service, e.g., for local area network (LAN) interconnection. Desired asymmetric VDSL data rates over 1 km are in the range 22–23 Mb/s downstream and 3–4 Mb/s upstream. This allows delivery of digital TV (DTV), high-definition TV (HDTV), multimedia entertainment, super-fast Web surfing, file transfer, and virtual office at home. For shorter distances of 300 m or less, the downstream data rate can be as high as 52 Mb/s, thus offering the possibility for a service provider to simultaneously deliver multiple DTV or HDTV channels. Symmetric

VDSL allows 13 Mb/s connections over twisted pairs of up to 1 km and 26 Mb/s over shorter loops of less than 300 m (Nava and Del-Toso [2002]).

As shown in Fig. 2.2, VDSL supports traditional narrowband POTS/ISDN services in the low-frequency spectrum of the same twisted pair. VDSL uses *frequency division duplex (FDD)* with several different frequency bands to separate upstream and downstream transmissions. There are two standards for single-carrier and multi-carrier modulation in VDSL. The multicarrier modulation is a DMT solution that is backward compatible with ADSL. The other solution uses QAM as a modulation scheme (Cioffi *et al.* [1999]). It is important to note, however, that only DMT remained part of the standard for second-generation VDSL systems, also known as VDSL2, which are able to provide up to 100 Mb/s in both upstream and downstream directions simultaneously for short local loops.

Recall from the above that tens or hundreds of twisted pair copper wires are bundled together in a single binder. Due to the electromagnetic coupling between twisted pairs the performance of xDSL systems is severely constrained. Many loop plants may carry different xDSL types that were potentially incompatible with existing services. For instance, in the United States, the 1996 Telecommunications Act created an environment to foster competition by unbundling the local loop of incumbent local exchange carriers (ILECs), thus allowing competitive local exchange carriers (CLECs) to provide xDSL services. As a consequence, ILECs could no longer control which xDSL technologies were deployed in their local loop plant. To allow xDSL deployments without incompatible technologies causing service outages, standardization efforts were initiated in 1998 to create a technical definition of spectral compatibility, giving rise to something that is now commonly referred to as *spectrum management*. Spectrum management is the process of ensuring spectral compatibility while optimizing the local loop plant. The resultant American National Standards Institute (ANSI) spectrum management standard T1.417-2001 defines a set of basis systems, which include HDSL, ADSL, and VDSL, among others. According to the spectrum management standard, any new xDSL technology must not generate crosstalk that significantly degrades the performance of any basis system. Toward this end, so-called *PSD masks* have been specified for the different xDSL technologies, whereby a given xDSL system cannot transmit more power at any frequency than the PSD mask to which it conforms in order to guarantee spectral compatibility (Kerpez [2002]).

The design of the PSD masks is based on worst case crosstalk scenarios, resulting in a static spectrum management and conservative xDSL deployments. Consequently, the current static design approach leads to xDSL deployments that are overly restrictive in terms of both coverage and speed. *Dynamic spectrum management (DSM)* departs from this design philosophy by allowing for adaptive allocation of the available resources among neighboring phone lines, depending on the channel characteristics and crosstalk levels (Song *et al.* [2002]). DSM DSLs coordinate simultaneous transmissions in the binder in order to reduce crosstalk and most other noises, e.g., radio ingress and impulse noise. They hold promise to use single-line VDSL systems to achieve 500 Mb/s data rates on phone lines of up to 400 m in length. It was shown that in multidropping DSM DSL architectures with two to four twisted pairs per customer, a binder of 200 phone line connections potentially provides a total of 100 Gb/s of shared bandwidth for both upstream and downstream directions together (Cioffi *et al.* [2007]).

While the data rates achievable in emerging DSM DSL architectures are beyond today's standardized optical fiber access network solutions, it is important to note that the speeds are higher not because copper has a wider bandwidth than fiber, but because existent optical access networks are far from exploiting the full bandwidth of fibers. Unlike DSM DSLs, optical access networks have not only the design goal of continuously increasing data rates but also, and more importantly, reducing the complexity, avoiding the use of powered equipment between CO and customers, as well as increasing the range while decreasing the CO_2 footprint for the realization of future-proof fixed wireline broadband access networks, as discussed in greater detail in Part II.

2.1.2 Cable modem

Cable television, originally known as community antenna television (CATV), started in 1948 with the goal to distribute analog TV programs in residential areas with limited reception of over-the-air broadcast TV signals. Traditional CATV networks consisted entirely of coaxial cables, which were typically laid out in a tree topology. At the root of the tree, the head-end receives the TV programs over satellite or microwave links. The head-end hosts the so-called *cable modem termination system (CMTS)*, which acts as an interface between the head-end and remote subscribers, each equipped with a *cable modem (CM)*.

Cable operators soon got the idea to utilize their network infrastructure to provide subscribers with not only television but also broadband Internet access and telephony services, turning traditional cable television operators into so-called multiple-system operators (MSOs) (Dutta-Roy [1999]). Conventional cable networks were required to have typically 20–40 amplifiers in cascade to compensate for the high propagation loss of coaxial cables. As an active device, each amplifier caused noise and nonlinear signal distortions and added to the unreliability of the amplifier chain. Due to the long cascade of amplifiers, early cable networks were not well suited for real-time two-way high-bandwidth services. As a result, beginning in the mid 1980s optical technologies were deployed by the cable industry to support broadband, giving rise to *hybrid fiber-coax (HFC)* architectures that benefit from significantly lower fiber losses and savings on the number of required amplifiers and their power supplies (Donaldson and Jones [2001]).

Figure 2.3 depicts the architecture of an HFC network. National TV programs are received from a geostationary satellite by a dish antenna at the head-end. The CMTS at the head-end mixes the national TV signals with news and commercials of local interest and transmits them in analog form to various neighborhoods by fiber optic cables, which terminate into fiber nodes, each supporting 500–2000 households. At the fiber nodes, the optical signals are converted into electronic signals and distributed over coaxial cable to the drop points of individual residences. From the drop point, a coaxial cable brings the signals to a network interface unit (NIU) located outside every residence. A diplexer

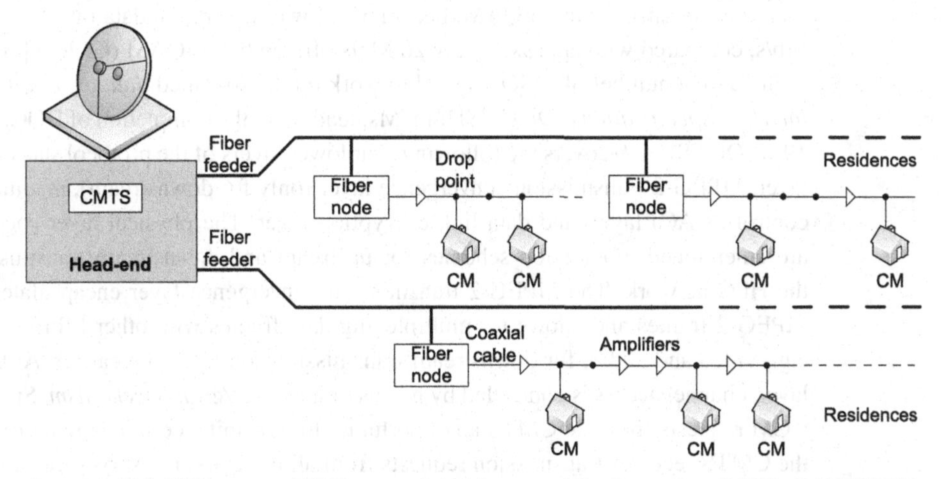

Figure 2.3 Hybrid fiber-coax (HFC) network architecture. After Dutta-Roy (1999). ©1999 IEEE.

inside the NIU separates the upstream from the downstream signals. The NIU also separates incoming signals destined for TV programs, voice telephony, and Internet data for the CM. Optical transmission of cable TV signals can be accomplished without any appreciable attenuation with single-mode optical fiber, which has an attenuation of approximately 0.35 dB/km at 1310 nm and 0.25 dB/km at 1550 nm. To compensate for the attenuation of lossy coaxial cables, bidirectional amplifiers are placed along the coaxial cables. To guarantee that all homes receive good quality signals, a home served by the CMTS is no more than some 100 km from the head-end. In the optical segment of the HFC architecture, signals from and to the head-end usually travel on separate fiber strands. In the coax segment, however, a single coaxial cable must carry both upstream and downstream signals. Toward this end, the downstream and upstream channels are carried on the common coaxial cable by using FDM with an assigned transmission spectrum that slightly varies for different regions. In the United States, individual TV channels are separated usually 6 MHz apart from each other, whereby downstream channels are placed at frequencies above 54 MHz, while frequencies between 5 MHz and 42 MHz are allocated to upstream channels for both analog and digital transmissions (Dutta-Roy [1999, 2001], Donaldson and Jones [2001]).

At each residence, the CM connects to the subscriber's personal computer through its Ethernet network interface card (NIC). Downstream data communication is accomplished with the same modulation systems used for cable digital television. There are two options, both using QAM in a 6-MHz channel. The slower system uses 64-QAM with an approximate raw data rate of 30 Mb/s, offering a payload information rate of 27 Mb/s after removing error correction and control overhead. The faster system uses 256-QAM with an approximate raw data rate of 43 Mb/s and a payload information rate of 39 Mb/s. A large number of sources of interference limits the upstream communication options and speeds due to the fact that signals leak into the cable system through consumer-owned devices and in-home wiring. Either 16-QAM or quadrature phase shift keying (QPSK) is used as modulation for upstream communication. QPSK can tolerate more interferences than 16-QAM, but offers a lower maximum data rate of roughly 10 Mb/s, compared with approximately 20 Mb/s offered by 16-QAM (Ciciora [2001]).

In 1995, a number of MSOs started to work on the so-called *data over cable service interface specifications (DOCSIS)* for CMs, leading to the completion of DOCSIS 1.0 in 1997. DOCSIS 1.0 covers the following four lower layers of the protocol stack: physical layer, MPEG-2 transmission convergence layer (only for downstream), medium access control (MAC) layer, and data link encryption layer. The physical layer specifies the aforementioned modulation schemes for upstream and downstream transmissions on the HFC network. The MPEG-2 transmission convergence layer encapsulates data in MPEG-2 frames and allows for multiplexing data frames with other MPEG-2 frames, e.g., video and audio, for downstream transmissions on the same carrier. At the MAC layer, channel access is controlled by means of a *request/grant mechanism*. Specifically, a CM requests from the CMTS an opportunity to transmit a certain amount of data. As the CMTS receives transmission requests from all the CMs, it reserves transmit opportunities on the upstream channel(s) and informs each CM about its respective transmit opportunity. In doing so, the CMTS is able to dynamically assign upstream transmit opportunities to all CMs. To guarantee collision-free upstream transmissions, the MAC protocol also defines procedures for the autodiscovery, ranging, synchronization, and channel acquisition of CMs. Finally, the data link encryption layer defines a simple data privacy function, known as baseline privacy interface (BPI), for encrypting traffic flows between each registered CM and the CMTS (Fellows and Jones [2001]).

Since the initial standardization in 1997, DOCSIS was revised several times to enhance its capabilities and functionalities. In 1999, DOCSIS 1.1 was released to advance the original best-effort service by providing guaranteed quality-of-service (QoS) to real-time applications, e.g., voice over Internet protocol (VoIP), to support IP multicast, and to offer enhanced security services. DOCSIS 2.0 was approved in 2001 to provide increased upstream data rates for symmetric services. The most recent version, DOCSIS 3.0, was released in 2006. DOCSIS 3.0 provides a number of enhancements, most notably, channel bonding, support for IP version 6 (IPv6), and creates a platform for the evolution of cable television into IP television (IPTV). With channel bonding, a single subscriber may use multiple upstream and/or downstream channels simultaneously, resulting in at least 120 Mb/s upstream and 160 Mb/s downstream (CableLabs [2010]).

It is worthwhile to note that the DOCSIS MAC protocol bears striking similarities to that of the Ethernet passive optical network (EPON), which will be described in greater detail in Chapter 4. In both DOCSIS-based HFC and EPON networks, upstream bandwidth is dynamically assigned according to given traffic demands by using a request/grant mechanism. Furthermore, both access networks deploy similar procedures for autodiscovery, ranging, and synchronization of customer premises equipment (CPE), albeit there are also some differences between the two systems. Among others, unlike EPON, cable networks allow for *fragmentation* in order to limit the size of upstream data frames. With fragmentation, the CMTS may instruct a given CM to fragment a large frame into multiple smaller subframes to ensure that the CMTS can schedule transmit opportunities between individual subframes for certain CMs when those opportunities are needed in order to maintain QoS guarantees (Fellows and Jones [2001]). More importantly, one of the major differences between HFC and EPON networks is the achievable data rate. As we will see in Chapter 4, state-of-the-art EPONs provide symmetrical and asymmetrical data rates of up to 10 Gb/s, which are way above those of DOCSIS 3.0 cable networks. In addition, EPONs bring fiber all the way to the end-user and help realize fiber-to-the-home/building (FTTH/B) networks. An interesting approach to offer a path for a seamless transition from cable-based services to FTTH services is the so-called *DOCSIS PON (DPON)*. In the past, MSOs have made significant and ongoing investments in operations and support system (OSS) software to integrate customer service, billing, and provisioning systems. The purpose of DPON is to provide a seamless integration of an MSO's existing OSS software with new EPON technologies into their commercial or residential offerings. By implementing the DOCSIS service layer interface on EPON, MSOs are able to provision and scale Ethernet-based business services, e.g., Ethernet LAN (ELAN), while maintaining backward compatibility (Blake [2008]).

2.1.3 Broadband over power line

Power line communications (PLC) have been studied for decades as a cost-efficient means to realize home automation, where existing power lines are used for in-home control and monitoring applications such as lighting control, security and access control, energy management, and automated meter reading (Bertsch [1990], Gershon *et al.* [1991]). Beside simple control and monitoring, PLC-based home networks provide an infrastructure for building *smart homes*, which seamlessly integrate many types of information appliances, e.g., multimedia entertainment systems, in support of real-time communications with QoS guarantees. Figure 2.4 depicts a smart home network, where one of the connected computers (in the upper-right corner of the figure) connects the in-home PLC network to the Internet using DSL or cable modem broadband technologies. As shown in Fig. 2.4, a smart home network may interconnect a

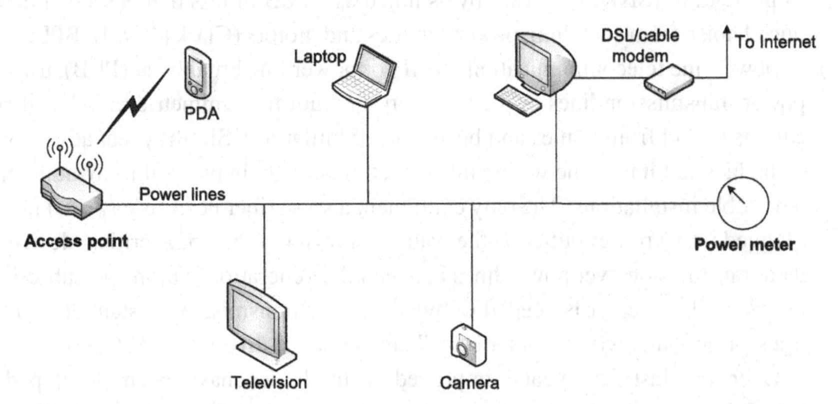

Figure 2.4 Smart home network using power line communications (PLC). After Lin *et al.* (2002). ©2002 IEEE.

wide range of different devices, including television, camera, personal digital assistant (PDA), wireless LAN access point, laptop computer, and power meter (Lin *et al.* [2002]).

The most important standard for PLC-based smart home networks is HomePlug AV (HPAV), which was developed by the HomePlug Powerline Alliance to achieve coexistence with legacy 50/60 Hz electricity as well as security and lighting control signals on household power wiring (HomePlug Powerline Alliance [2010]). HPAV's spectrum ranges from approximately 1.8 MHz to 30 MHz and can be easily separated from legacy narrowband signals by means of passive filtering. It uses variable-bit loading orthogonal frequency division multiplexing (OFDM), where selected carriers may be turned off to mitigate induced electromagnetic or radio frequency interference (EMI/RFI). More specifically, with variable-bit loading OFDM, the HPAV spectrum is divided into multiple equally spaced carriers and each carrier is dynamically loaded with data bits according to given channel interference and other line conditions. Depending on channel conditions, each carrier is modulated using either binary phase shift keying (BPSK), QPSK, 8-QAM, 16-QAM, 64-QAM, 256-QAM, or 1024-QAM, providing a physical layer data rate of up to roughly 200 Mb/s (Hazen [2008]).

PLC technologies may be used in two areas: (*i*) in-home networking inside the home, which is sometimes referred to as *last inch* access, and (*ii*) *last mile* access to the home. While PLC could be the preferred last-inch access technology of choice, it is not widely thought to be superior to other last-mile broadband access technologies such as DSL and cable modem, except for rural, remote, or underserved areas where telephone and cable connections may not exist (Majumder and Caffery [2004]). A recent example of deploying broadband over power lines is the region of Larissa, a rural area in central Greece, where broadband access is provided along a power grid (in conjunction with wireless LAN access points) (Sarafi *et al.* [2009]).

Broadband over power line (BPL) last-mile access networks face much harsher transmission conditions than their DSL and HFC based counterparts. Unlike in twisted pair copper and coax based architectures, frequency-dependent attenuation can vary by as much as 20 dB in less than a second by turning on appliances. Furthermore, the signal-to-noise ratio (SNR) may vary by as much as 10 dB in less than a second due to introduced noise from switching power supplies and motors (Clark [1998]). BPL, also known as power line telecommunications (PLT) or power line broadband (PLB), uses electrical power transmission lines as the transport medium for symmetric broadband communications to and from homes and businesses. Similar to DSL, the great advantage of BPL is the fact that it uses the wiring infrastructure already in place, thus avoiding the cost of new cable installations. The only equipment a subscriber needs is a special modem to be plugged into a power outlet on the wall, as shown in Fig. 2.5. A critical device to enable data transmission over power lines between the generation station and subscribers is the *coupler*. The coupler is needed to bypass power transmission system elements such as transformers that would hinder broadband signals delivery (Qiu [2007]).

Over the last few years, improved technologies have been developed to send broadband signals along with electricity in last-mile BPL access networks. Recently, research on ultra-wideband (UWB) communication over indoor power lines has started,

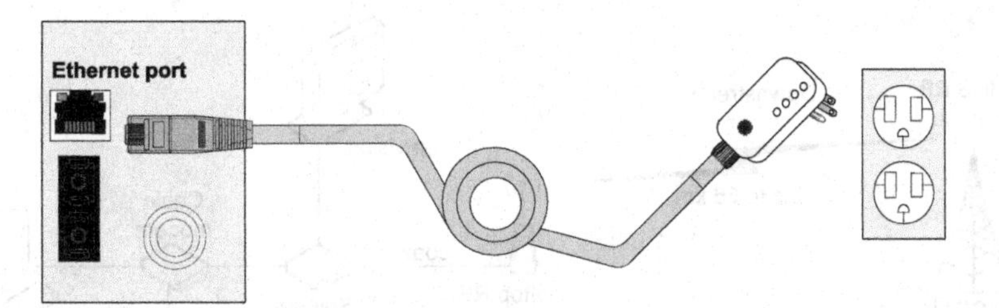

Figure 2.5 Broadband over power line (BPL) modem with Ethernet connection to the computer. After Qiu (2007). ©2007 IEEE.

exploiting frequencies above 30 MHz. Preliminary studies indicate that indoor PLC systems with Gigabit per second data rates become feasible by using a frequency range from 50 MHz to 550 MHz (Chen *et al.* [2009]).

2.2 Fixed wireless broadband technologies

2.2.1 MMDS

Multichannel multipoint distribution service (MMDS) is a line-of-sight (LOS) wireless technology that uses the frequency range of 2.150 to 2.680 GHz. MMDS simultaneously applies multiple channels to support various traffic types. In the early 1970s, the Federal Communications Commission (FCC) allocated the first two channels (i.e., 2.150 and 2.160 GHz) to business data transmissions. Due to the fact that the FCC did not specify the types of traffic, service providers applied one of them for video and another one for voice and data traffic (Biagini and Yang [2000]). MMDS was originally designed to broadcast TV over a single cell with up to 56 km radius in a point-to-multipoint manner. Since the MMDS technology is similar to cable television, it is also known as wireless cable. Due to the inherent property of MMDS of providing wireless service to both business and home end-users in various terrain types and especially in rural areas, where cable installation is difficult or costly, it was considered a promising technology to provide not only TV but also fixed wireless broadband access for Internet end-users. In both MMDS and HFC networks, the medium is shared among end-users using radio frequency (RF) signals to carry data.

As shown in Fig. 2.6, the MMDS RF channel is used in the downstream direction, while cable modem and coax cable are used in the upstream direction. In an MMDS network, the end-user is equipped with a DOCSIS cable modem. Typically, the MMDS RF link provides up to 2 Mb/s. We note that the capacity of an MMDS network can be further increased by means of RF multiplexing techniques or defining multiple sector cells. In order to apply MMDS with sufficient bandwidth in a large metropolitan area, cellular MMDS architectures should be designed, where the radio spectrum can be reused. AT&T labs proposed and examined a two-way MMDS broadband fixed wireless access network, where both upstream and downstream transmissions are done wirelessly (Kim *et al.* [1999]). In the proposed architecture, the frequency of 2.6 GHz was

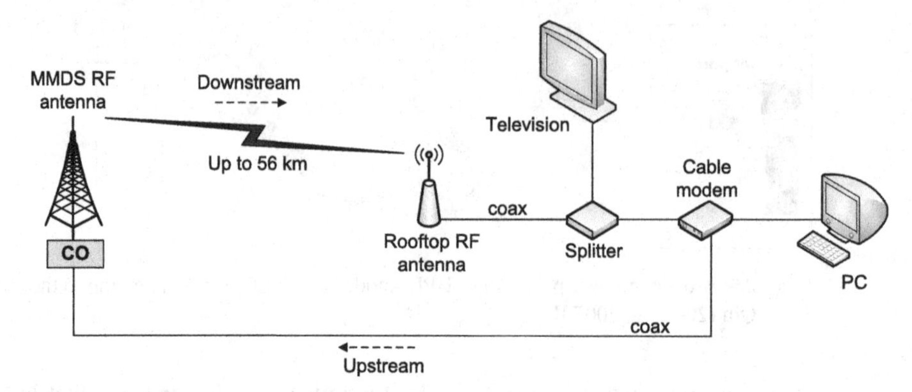

Figure 2.6 MMDS network architecture.

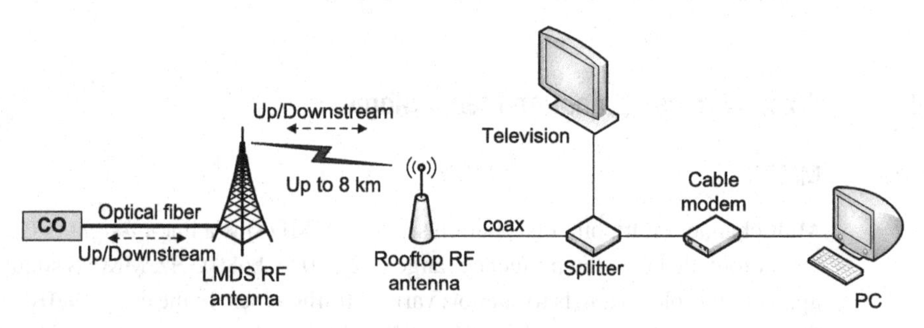

Figure 2.7 LMDS network architecture.

considered for realizing a downstream data rate of 27 Mb/s, while on the upstream link the frequency was set to 2.1 GHz providing a data rate of 2 Mb/s.

Local multipoint distribution system (LMDS) is another LOS-based fixed wireless broadband access technology, which applies the frequency ranges of 27.5 to 29.5 GHz and 31.0 to 31.3 GHz. Figure 2.7 shows the LMDS network architecture. LMDS is a two-way wireless technology that provides up to 155 Mb/s downstream and 1.54 Mb/s upstream data rates. Although LMDS offers much higher data rates than MMDS, the range of its radio signals is limited to approximately 8 km due to large free space attenuation rendering it a cost-efficient approach for wireless end-users in urban terrain type.

Since both LMDS and MMDS are LOS-based wireless technologies, various terrain obstructions such as mountains, trees, and buildings have a major impact on their coverage and signal quality. Foliated trees may attenuate the MMDS and LMDS RF signals significantly (Papazian *et al.* [1997]). Moreover, the movement of tree branches in windy conditions results in large and rapid signal fading. According to (Pelet *et al.* [2004]), winds of 15 km/h cause fades of 15 dB for an MMDS RF signal destined to the antenna on a bluff of four poplar trees. While rainy conditions have no major impact on the quality of the MMDS RF signal, the increased scattering caused by wet leaves increases signal fading (Kim *et al.* [1999]).

2.2.2 Free space optics

Free space optics (FSO), also known as optical wireless communications, was already tested in the 1960s, well before the development of optical fiber communications systems (Davis *et al.* [2003]). Unlike optical fiber systems, FSO systems may be installed on a demand basis without requiring the construction of expensive cabling infrastructures and may be deployed to connect neighboring business buildings without any right-of-way. They operate at high carrier frequencies and are thus able to provide multi-Gb/s LOS wireless point-to-point communications links of typically a few hundreds of meters. As the RF spectrum is becoming increasingly oversubscribed, FSO systems represent an interesting alternative that avoids the spectrum and data rate limitations of RF systems by using much higher frequency bands, ranging from 60 GHz up to optical frequencies of almost 200 THz. FSO systems deploy a laser to generate a highly directional lightwave that carries data through the atmosphere to an optical receiver located at a fixed distance from the transmitter. FSO links are commercially available that offer 1, 10, or even 40 Gb/s transmission rates over short distances (Wells [2009]).

A key challenge in FSO communications systems is a process known as *pointing, acquisition, and tracking (PAT)* for establishing and maintaining an optical link between two sites (Epple and Henniger [2007]). PAT involves a number of steps. The first step is to point the FSO terminal with its telescope toward the intended communication partner based on available information about the partner's position. If the terminal is unable to determine the position, it has to do a wide scan for the partner terminal. This task is in most cases time consuming and, if both terminals have to do scanning, it can become very complex. After completing the pointing step, an optical signal from the partner should be visible, which can be used to implement a closed-loop control system to correct the initial pointing and align the two terminals more precisely. After the terminals have acquired each other, the FSO system is ready for communication. To maintain the communication, each terminal tracks the optical signal from the partner and aligns itself with it. One of the biggest challenges for widespread deployment of FSO systems is their optical signal propagation in different atmospheric conditions, which may cause hugely varying link degradation due to variable attenuation and fading. Atmospheric turbulence creates temporary pockets of air with different refractive indices. These pockets let the laser beam phase-front vary randomly, resulting in intensity fluctuation.

Despite recent progress in indoor optical–wireless systems (Green *et al.* [2008]), outdoor deployment of FSO systems has been hindered by their lack of carrier-class availability in the presence of atmospheric variations. According to Nadeem *et al.* (Nadeem *et al.* [2009]), optical attenuation can reach up to 120 dB/km in moderate continental fog environments in winter months and 480 dB/km in dense maritime fog environments in summer months. Such conditions reduce the availability of FSO links dramatically. To mitigate the detrimental impact of weather effects, FSO links can be backed up with RF links to achieve near carrier-class availability. Table 2.1 compares the availability of an FSO link, a 40 GHz RF link, and a hybrid FSO/RF link based on measurements for dense fog, rain, and snow. The table clearly shows

Table 2.1. Availability of FSO, RF, and hybrid FSO/RF systems (Nadeem *et al.* [2009]).

	FSO	RF	Hybrid FSO/RF
Dense fog	0.51%	100%	100%
Rain	85.71%	14.29%	85.71%
Snow	39.49%	100%	100%

that using a combination of FSO primary link and RF backup link is a viable solution to alleviate the impact of varying weather conditions and improve the link availability significantly. Hybrid FSO/RF links combine the reliability of RF links with the high capacity of optical–wireless links and are suited for both commercial and military applications (Nadeem *et al.* [2009], Stotts *et al.* [2009]).

2.2.3 Satellite

Beside the above discussed terrestrial networks, satellite networks offer an interesting alternative for fixed wireless broadband access. Satellite communications networks have been successfully deployed for decades (Campanella and Harrington [1984]). As opposed to HFC networks, satellite networks provide consumers with the opportunity to enjoy *direct-to-home (DTH)* digital television without requiring any wired distribution network infrastructure (Dulac and Godwin [2006]). Satellite networks may provide broadband access globally on the entire surface of the Earth or geographically limited access to remote areas that don't have any terrestrial network infrastructure. Depending on their orbit, satellites may be classified into the following three major categories:

- **Geostationary orbit (GSO)**
  The majority of deployed communications satellites are GSO satellites. The circular orbit altitude of a GSO satellite is 35 786 km above the equator and the satellite's movement is synchronized with the rotation of the Earth. The round-trip propagation delay to and from a GSO satellite is in the range 250–280 ms. Three GSO satellites are sufficient for global coverage.

- **Medium Earth orbit (MEO)**
  MEO satellites move at a lower orbit than GSO satellites. MEO satellites can be as close as 3000 km from the Earth's surface and have a typical round-trip propagation delay of 110–130 ms.

- **Low Earth orbit (LEO)**
  LEO satellites travel at even smaller distances from the Earth's surface than their MEO counterparts. The round-trip propagation delay of an LEO satellite is 20–25 ms, which is comparable to that of terrestrial links.

Compared with GSO satellites, MEO and LEO satellites require smaller antennas and transmission power levels due to the fact that they are closer to the Earth. On the downside, a large number of MEO/LEO satellites is needed to achieve global coverage. Furthermore, users may need to be handed off from satellite to satellite as MEO/LEO satellites move rapidly over the surface of the Earth. The most commonly used frequency bands for satellite communications are the C band (4–8 GHz), Ku band (10–18 GHz), and Ka band (18–31 GHz) (Hu and Li [2001]).

According to Bem *et al.* [2000], satellite systems can act as broadband access or core networks. Figure 2.8 depicts a satellite broadband access network, where the satellite receives the signal from a user's terminal and retransmits it to a gateway, which in turn forwards the received signal to the service provider via a terrestrial core network. The satellite acts as a simple repeater between the user's terminal and the gateway. In a satellite broadband access and core network, the signal received from the user's terminal is sent by the satellite across *intersatellite links (ISLs)* of a high-speed satellite core network that runs in parallel to the terrestrial core network, as shown in Fig. 2.9. In this case, satellites are equipped with *onboard processing (OBP)* systems for advanced modulation, coding, switching, and routing operations in support of multimedia applications (Wittig [2000]). ISLs may be used to form satellite core networks that operate at optical frequencies and offer data rates of >10 Gb/s. These optical satellite networks

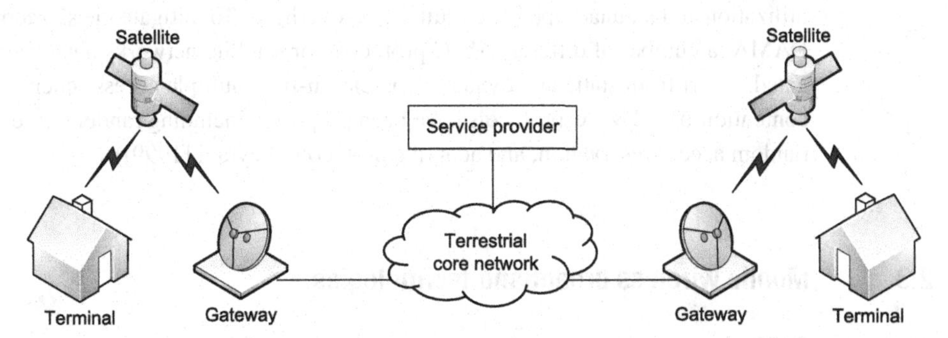

Figure 2.8 Satellite broadband access network. After Bem *et al.* (2000). ©2000 IEEE.

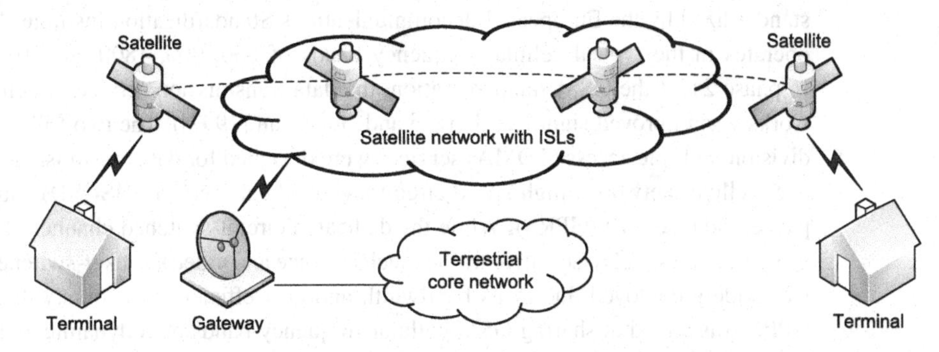

Figure 2.9 Satellite broadband access and core network with intersatellite links (ISLs). After Bem *et al.* (2000). ©2000 IEEE.

may represent a cost-competitive alternative to undersea fiber systems and terrestrial core networks. Optical satellite networks may serve as an alternate backbone to restore global connectivity when natural or man-made disasters disconnect parts of terrestrial access or core networks (Chan [2003]).

Satellite networks are inherently broadcasting systems, providing point-to-multipoint communications in the downstream direction (from satellite to users). Accordingly, they have been traditionally used for the unidirectional distribution of broadcast television signals. Due to the development of return-channel technologies for satellite-based multimedia communications, broadband satellite access systems for bidirectional interactive multimedia Internet services have become possible, giving rise to so-called *broadband satellite access (BSA)* systems (Le-Ngoc *et al.* [2003]). Typically, demand assigned multiple access (DAMA) schemes are deployed in satellite communications systems for dynamic capacity allocation. DAMA schemes make use of a request/grant mechanism to set up and tear down circuits between a terminal and the satellite network control station, which is usually ground based. The establishment of circuits via a satellite is a two-way reservation process and implies twice the round-trip propagation delay to and from the satellite, amounting to a total of up to 560 ms in the case of GSO satellites. Clearly, DAMA schemes provide efficient dynamic capacity allocation for nonbursty traffic with relatively long holding times, e.g., voice traffic. For bursty data traffic and/or short messages, however, DAMA schemes suffer from a poor bandwidth utilization and an unacceptable circuit set-up overhead. To mitigate the shortcomings of DAMA, a number of different MAC protocols for satellite networks have been investigated. Apart from static and dynamic contention-free multiple access schemes, various contention-based MAC protocols have been proposed, including random access, hybrid random access/reservation, and adaptive protocols (Peyravi [1999]).

2.3 Mobile wireless broadband technologies

2.3.1 GPRS

The global system for mobile communications (GSM) is a digital cellular technology standardized by the European Telecommunications Standardization Institute (ETSI). It operates in the typical cellular frequency bands of 850, 900, 1800, and 1900 MHz. In phase 2+ of the GSM standardization, the data transmission service of cellular networks was improved significantly (Cai and Goodman [1997]). The two following time division multiple access (TDMA) services were designed for data transmission services over cellular networks: high-speed circuit-switched data service (HSCSD) and general packet radio service (GPRS). While the dedicated circuit-switched channel of HSCSD was suitable for delay-sensitive traffic (such as voice and video), packet-switched GPRS was widely deployed due to its bandwidth and cost efficiency for bursty data traffic. GPRS was aimed at sharing GSM cellular frequency bands on a dynamic and flexible basis to provide both the packet data service of GPRS and the circuit-switched service of GSM, such as voice and short message service (SMS). While GSM provides digital

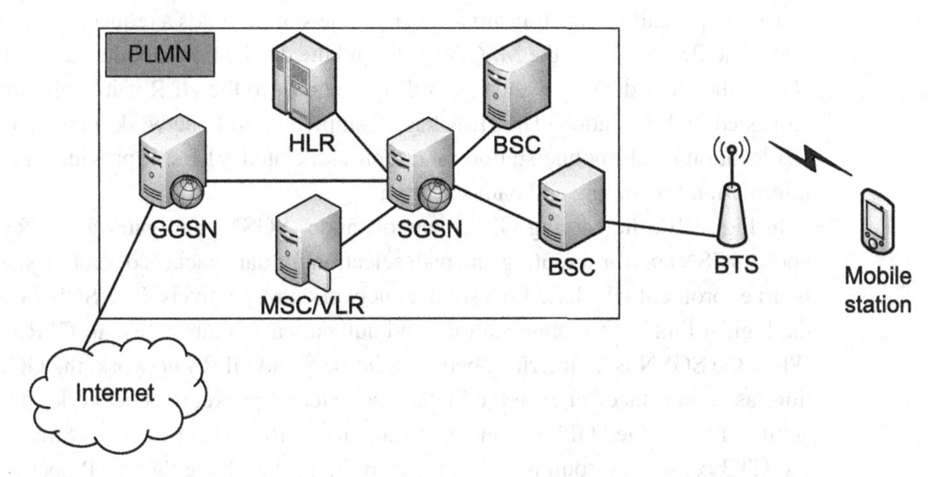

Figure 2.10 GPRS network architecture. After Sarikaya (2000). ©2000 IEEE.

data interchange of only up to 9.6 kb/s when the cellular phone is used as a modem, GPRS provides maximum data rates of 165.5 kb/s and 154 kb/s for downstream and upstream directions, respectively (Sarikaya [2000]).

Figure 2.10 shows the GPRS network architecture. In this figure, a base station controller (BSC) controls the base transceiver station (BTS). The BTS is equipped with an antenna to provide service to the mobile stations in its radio coverage (cell). In GPRS networks, multiple BTSs might be connected to a single BSC. A BSC and its connected BTSs form a basic station system (BSS). In Fig. 2.10, the GSM public land mobile network (PLMN) comprises the entire service area (including sets of cells) of a network operator. The PLMN connects the mobile stations to the wired public switched telephone network (PSTN). It consists of the mobile switching center (MSC), which monitors and controls one or more BSC. The MSC performs the telephony switching and authenticates the access of mobile stations to GSM circuit-switched services (Samjani [2002]). The service profile of each mobile station and its last known location information are stored in the home location register (HLR). The visiting location register (VLR) registers new visiting mobile stations and temporarily stores their location information. VLR and HLR are databases that are used to realize handovers between two cells (Sarikaya [2000]).

When a mobile station is initially turned on it will be registered at the VLR. The MSC checks the credit validity and service profile of the connected mobile station at the HLR. Once the access of a mobile station is authenticated, the HLR keeps the location of the mobile station. Typically, the BSC in a GPRS network decides to perform handover for each mobile station depending on various parameters, such as signal strength. The handover can be intra-BSS or inter-BSS, where the latter can be intra-MSC or inter-MSC. Although the MSC/VLR must be involved in inter-BSS handovers, the BSC makes the final decision of performing a handover since the MSC has no real-time information about the connection status. For inter-MSC handover, the same procedure as in registration is carried out to transfer a connection to another MSC/VLR.

To set up a call or data transmission, a mobile station sends a request to its associated BTS. The BSC requests the MSC/VLR to find the destination mobile station. The MSC checks the VLR data base and forwards the request to the HLR if it could not find the requested mobile station. The HLR has a complete mobile network map, including the last location of all mobile stations and their associated VLRs. It provides the required information for routing and path selection.

In Fig. 2.10, the serving GPRS support node (SGSN) and gateway GPRS support node (GGSN) perform routing and path selection for data packets of mobile stations and Internet protocol (IP)-based packet data networks, respectively. The SGSN establishes the logical link with mobile stations and authenticates their access to GPRS services. While the SGSN is an interface between the BSS and GPRS network, the GGSN functions as an interface between the SGSN and external packet data networks. The GGSN applies the so-called GPRS tunneling protocol (GTP) to encapsulate and transmit packets. GTP adds some routing information to the packet. Note that GTP operates on top of the transport and network layers. Various Ethernet- and asynchronous transfer mode (ATM)-based protocols may be deployed in conjunction with GTP (Cai and Goodman [1997]). It is worthwhile to mention that inter-PLMN traffic is transferred using private IP addresses (either IPv4 or IPv6), while a roaming agreement installed between two GPRS networks is used for intra-PLMN traffic (Bettstetter *et al.* [1999]).

To increase the service coverage and data rate in buildings, the integration of GPRS and widely deployed wireless local area network (WLAN) was proposed in (Chen and Chen [2007]). To simplify handovers, a centralized mobility management node was introduced. In this heterogeneous wireless network, the mobility of end-users is supported by using mobile IP (MIP), which was standardized in Internet Engineering Task Force (IETF) RFC 3344 and RFC 3775 for IPv4 and IPv6, respectively.

2.3.2 EDGE

Enhanced data rates for GSM evolution (EDGE) was introduced by ETSI in 1997 to increase user bit rates and improve spectral efficiency. EDGE is designed in a pay-as-you-grow way using the deployed infrastructure and frequency of GSM and GPRS networks. More specifically, EDGE provides an evolutionary path from existing cellular network standards to third-generation services over the existing spectrum bands (Furuskär *et al.* [1999]). Table 2.2 shows the cellular network technology evolution from 2000 to 2013. The table shows that EDGE is expected to play a key role in future cellular networks.

In an EDGE cellular network, the GSM carrier signal is used for high bit rate data transmissions. This backward compatibility of EDGE with legacy cellular systems renders EDGE technology a narrowband International Mobile Telecommunications (IMT) option, also known as IMT single-carrier (SC) (Callendar [2010]). The first EDGE network was deployed by AT&T in 2003 and has been deployed widely during the last decade.

As shown in Fig. 2.11, the EDGE standardization included two phases to provide different services (Molkdar *et al.* [2002]): (*i*) enhanced general packet radio service

Table 2.2. Number of mobile subscriptions (in thousands) in cellular network technology evolution (Callendar [2010]).

	2G	W-CDMA	CDMA2000	EDGE	Total
2000	800 000	0	0	0	800 000
2001	950 000	0	0	0	950 000
2002	1 050 000	0	50 000	0	1 100 000
2003	1 200 000	0	100 000	0	1 300 000
2004	1 600 000	0	200 000	0	1 800 000
2005	1 850 000	50 000	250 000	50 000	2 200 000
2006	2 200 000	100 000	300 000	100 000	2 700 000
2007	2 450 000	200 000	350 000	300 000	3 300 000
2008	2 150 000	350 000	500 000	800 000	3 800 000
2009	1 800 000	500 000	550 000	1 200 000	4 050 000
2010	1 550 000	600 000	550 000	1 500 000	4 200 000
2011	1 350 000	850 000	600 000	1 700 000	4 500 000
2012	1 350 000	1 000 000	550 000	1 800 000	4 700 000
2013	950 000	1 400 000	500 000	2 000 000	4 850 000

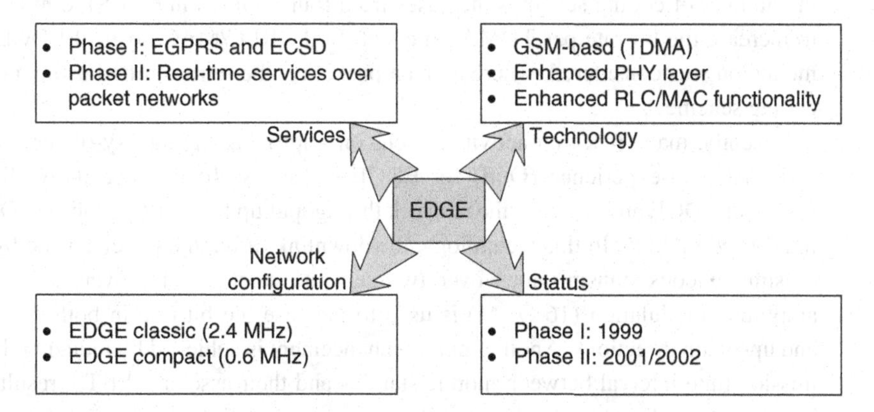

Figure 2.11 High-level view of EDGE. After Molkdar *et al.* (2002). ©2002 IEEE.

(EGPRS) and enhanced circuit switched data (ECSD) to enhance packet-switched and circuit-switched data services, respectively; and (ii) real-time services (such as video telephony) over packet-based networks.

Various techniques were proposed to enhance the physical (PHY) and radio link control (RLC)/medium access control (MAC) layers of an EDGE network, while preserving the main fundamental physical characteristics of GSM and GPRS networks. As illustrated in Fig. 2.11, the following two air interface configurations were proposed to allow GSM operators to deploy EDGE with their already deployed equipment: (i) EDGE classic applies 2.4 MHz channels, and (ii) EDGE compact applies 0.6 MHz channels designed for countries without any GSM infrastructure or where spectrum is scarce.

Table 2.3. EDGE high-level specifications (Molkdar *et al.* [2002]).

	Indoor/Outdoor	Urban/Suburban outdoor	Rural outdoor
User data rate (kb/s per time slot)	48	48	18
User data rate (kb/s per carrier)	384	384	384
User speed (km/h)	up to 10	100	250
Radio condition	Indoor and TU3	TU50 and HT100	RA250 (900 MHz) RA130 (1800 MHz)

Table 2.3 summarizes the EDGE high-level specifications. In addition to the legacy Gaussian minimum shift keying (GMSK) modulation technique of GSM, 8-ary phase shift keying (8-PSK) modulation was introduced to improve the spectral efficiency of EDGE. Incremental redundancy (IR) is a code combining the technique proposed in the EDGE standard in order to provide hybrid automatic repeat request (ARQ). In EGPRS, the number of coding schemes increases from four (defined in GPRS) to nine in order to increase the bit rate per TDMA time slot. In the RLC/MAC layer of EDGE, resegmentation was considered, where a given packet can be retransmitted using a different coding scheme.

Recently, many research activities focus on improving the quality-of-service (QoS) and quality-of-experience (QoE) for EDGE end-users. To improve the QoS performance of EDGE and increase the network throughput up to 1 Mb/s, evolved EDGE was introduced in 2006. In this technology, the downlink throughput is enhanced by means of simultaneous transmissions over two separate carriers. Moreover, 16-quadrature amplitude modulation (16-QAM) is used to increase the bit rate in both downstream and upstream directions. Another major enhancement is achieved by reducing the transmission time interval between mobile stations and their associated BTS, resulting in a significantly reduced latency.

The evolved EDGE provides various enhancements in different layers:

- Spectrum efficiency/capacity: up to 50% in kb/s/MHz/cell for data and Erl/MHz/cell for speech.
- Increase of peak data rates: up to 100% in both downstream and upstream directions.
- Improved coverage: increased sensitivity for speech and data in downlink for up to 3 dB.
- Improved service availability: up to 50% mean bit rate increase at cell edges for both downlink and uplink.
- Reduced latency: call setup and initial access is performed within a round-trip time (RTT) of less than 450 ms, while service is provided within an RTT of less than 100 ms.

- Balanced performance improvement: downlink and uplink throughput improvement under realistic network conditions in terms of bit rate, coverage, capacity, and latency.

Applying the above-mentioned enhancements, evolved EDGE aims at providing different services and supporting various applications: (i) interactive and best-effort services such as web browsing, which benefit from the increased mean bit rate and reduced latency; (ii) conversational services such as voice over Internet protocol (VoIP), enhanced push to talk over cellular (PoC), and on line gaming services, which require small latency with quick access; (iii) video telephony service is supported using an increased coverage for higher bit rates in both uplink and downlink; (iv) TV broadcasting, which typically needs high bit rates, low latency, and robust coverage.

One important issue in the design of the evolved EDGE technology is the backward compatibility with existing legacy frequency planning. This enables operators to deploy the proposed enhancements of evolved EDGE in their existing networks while keeping the same level of frequency protection, sensitivity, and interference. The compatibility with legacy EGPRS mobile stations is another issue that can be achieved by allowing the multiplexing of shared resources. One of the major compatibility objectives of the evolved EDGE technology is to avoid any changes on the hardware of legacy EDGE equipment such as BTS and BSC.

2.3.3 UMTS

The universal mobile telecommunications system (UMTS), also known as the third-generation mobile communication system, was designed by ETSI to improve the QoS support of mobile end-users. One of the important parameters in the design and standardization of UMTS was the market requirements gathered by the UMTS forum and GSM memorandum of understanding (MoU) association (Samukic [1998]).

In 1999, the initial UMTS network was specified (3GPP [2010b]). Although UMTS is based on the concepts of GSM and EDGE networks, it is important to note that UMTS deployments require a new radio frequency allocation and major hardware modifications of service provider equipment, i.e., base stations (BSs). To support legacy GSM and EDGE networks, UMTS mobile handsets are designed in a way that allows seamless dual-mode operation. For different indoor, urban/suburban, and rural terrain types, UMTS provides data rates of up to 2 Mb/s, 384 kb/s, and 144 kb/s, respectively (Richardson [2000]).

Figure 2.12 shows the UMTS network architecture, where the mobile stations are connected to the core network (CN) through a radio access network (RAN) (Moustafa *et al.* [2002]). In this architecture, a gateway node (GWN) connects the UMTS network to the Internet, PSTN, and legacy GSM mobile networks. As shown in Fig. 2.12, an RAN is responsible for all radio frequency allocations and includes a set of interconnected radio network systems (RNSs). In an RNS, the channel resources of its attached BSs are managed and controlled by the radio network controller (RNC). Moreover, the intercell handovers are also performed and controlled by the RNC.

Table 2.4. UMTS traffic classes (Moustafa *et al.* [2002]).

		Conventional	Streaming	Interactive	Background
Feature	Application	VoIP	Audio/Video	Web browsing	Email
	Delay	Strict and low	Bounded	Tolerable	Unbounded
	BER	10^{-3}	10^{-5}	10^{-8}	10^{-8}
Attribute	Maximum bit rate	Yes	Yes	Yes	Yes
	Guaranteed bit rate	Yes	Yes	No	No
	Maximum delay	Yes	Yes	No	No

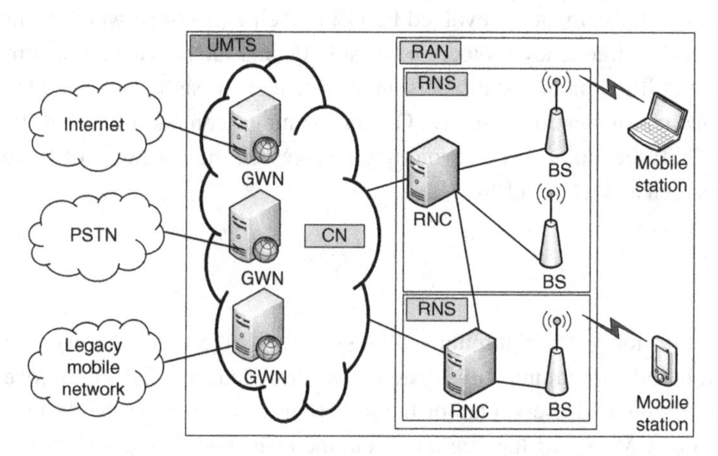

Figure 2.12 UMTS network architecture. After Moustafa *et al.* (2002). ©2002 IEEE.

One of the important challenging issues in UMTS networks is the QoS support for various types of traffic. Figure 2.13 shows the radio resource management module of BS nodes in the UMTS architecture (Jorguseski *et al.* [2001]). Connection admission control is the initial operation performed for radio resource assignment. After the session requests are accepted by the admission control module, they are classified in different queues based on their priorities. Table 2.4 summarizes the fundamental features and attributes of the four traffic classes defined in the UMTS standard (3GPP [2010b]).

In Fig. 2.13, the time, power, and rate schedulers are in charge of scheduling the requests according to predefined scheduling algorithms. The resource estimator synchronizes and controls all resource management modules based on measured interference conditions, radio channel characteristics, current traffic loads in the BS, and QoS requirements. More specifically, the resource estimator performs the following three tasks:

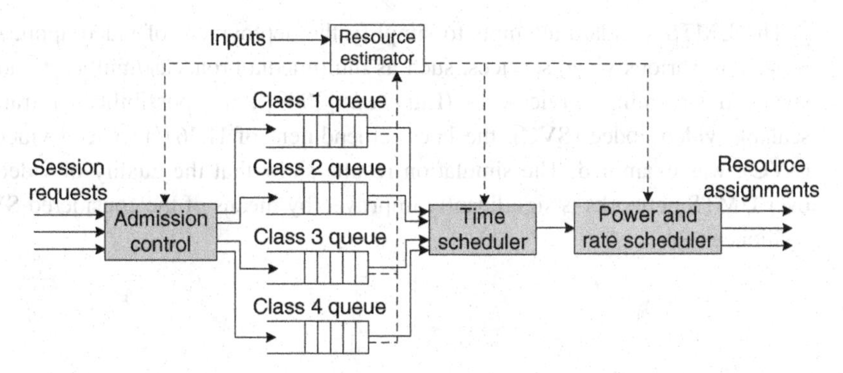

Figure 2.13 Radio resource allocation in UMTS network. After Jorguseski *et al.* (2001). ©2001 IEEE.

1. It applies the radio channel characteristics and session QoS requirements for optimal power and rate allocation.
2. It uses the current BS load (status of queues), session traffic characteristics, and session QoS requirements to control the time scheduler.
3. By monitoring all resource management modules, it assists the admission control accepting or declining the incoming session requests.

During the last decade, various amendments have been released. The latest release 10 (3GPP [2010a]) studies the network performance improvement techniques and smooth upgrades to fourth-generation mobile communication systems, also known as long term evolution (LTE). This amendment studies different modifications ranging from node architectures to management schemes, e.g., energy management.

Very recently, dual-mode handheld devices have been widely released to operate with WLAN and UMTS networks either simultaneously or interchangeably. The handover between WLAN and UMTS networks represents an interesting means to increase the QoS level of the provided service. The vertical handover is an approach proposed in (Liao and Chen [2008]), where a mobile station at different instants is associated with the antennas of a suitable network. In this scheme, a mobile station initially connects to a WLAN antenna due to its lower cost and higher bandwidth and attempts to connect to a UMTS network if the received signal strength from the WLAN changes dramatically. In the proposed technique, various challenging issues such as low handover latency are addressed for a seamless vertical handover without any modifications of WLAN and UMTS network equipment, whereby only edge router nodes of both networks should be upgraded to support vertical handovers.

Another recently proposed research area is the design of several UMTS multi-antenna systems with radiators to increase isolation and decrease envelope correlation (Diallo *et al.* [2008]). Using the presented optimization techniques, the isolation between the antennas is increased to enhance their efficiency. Moreover, the correlation of the received signals is reduced, which is crucial to ensuring improved diversity and MIMO performance.

The UMTS standard attempts to simplify the deployment of video applications by proposing various video services, such as multimedia broadcast/multicast and packet-switched streaming services. In (Liu *et al.* [2009]), the possibility of transmitting scalable video codec (SVC), the latest amendment of H.264/advanced video coding (AVC), was examined. The simulation results show that the quality of video service over UMTS networks is significantly improved by means of the considered SVC error resilience tool.

Part II

Fiber access networks

Fiber access networks have in general one of the following three architectures: (i) point-to-point architecture, (ii) active star architecture, or (iii) passive star architecture. In the point-to-point architecture, each home or building is connected to the central office (CO) via one or two dedicated fibers. This type of architecture provides improved privacy and ease of service upgrade for individual subscribers, but requires a large number of fibers and transceivers since network equipment is not shared among subscribers. As a consequence, footprint and power consumption may become serious problems at the CO. This shortcoming is avoided in star architectures, where a single feeder fiber runs from the CO to a remote node, from which individual distribution fibers branch out to connect the subscribers. The feeder fiber carries all the traffic of the attached subscribers and its cost can be shared among them. In doing so, the number of required fibers and transceivers at the CO can be reduced significantly. Depending on the nature of the remote node, the star architecture may be either active or passive. In the active star architecture, the remote node is an active device, e.g., Ethernet switch, and needs powering and maintenance. Conversely, in the passive star architecture, the active node is replaced with a passive optical splitter/combiner. Using a completely passive splitter/combiner at the remote node avoids the need for powering and maintenance and thereby helps reduce the capital expenditures (CAPEX) and in particular operational expenditures (OPEX) of fiber access networks (Koonen [2006]).

Fiber access networks have been envisioned for delivering broadband services for over 30 years. However, per-subscriber costs of early point-to-point field trials were far higher than existent alternative broadband solutions. Redesigning fiber access networks from a dedicated point-to-point architecture to a point-to-multipoint passive star architecture made it possible to achieve the necessary cost savings for the wide deployment of so-called *passive optical networks (PONs)*. Among others, PONs exhibit the following highly desirable properties (Shumate [2008]):

- No need for active outside equipment and backup batteries.
- No ongoing utility costs associated with the power consumption of active devices.
- No issues of electromagnetic interference (EMI) or electromagnetic compatibility (EMC).
- Reduced network failure rate and associated repair costs.
- No use of bandwidth-dependent technology and thus substantially increased options for future upgrades.

Due to their completely passive nature, PONs incur lower CAPEX and OPEX and also offer a higher reliability than active star architectures. Furthermore, PON outside plants provide transparency against data rate, modulation format, and protocol as the passive splitter/combiner is entirely agnostic to all three of them. This transparency, apart from the huge bandwidth and low loss of optical fiber, is one of the most crucial features that eased carriers into deploying PON-based fiber access networks, which are instrumental in minimizing deployment costs while maximizing revenues from new service offerings and can be flexibly upgraded as new technologies mature or new standards evolve (Effenberger *et al.* [2007]).

PONs deploy a power splitter/combiner at the remote node to equally divide downstream light among its N distribution fibers and to combine upstream light arriving on them, thereby experiencing a splitting and combining loss equal to 10 log N in the downstream and upstream direction, respectively. Typically, upstream and downstream transmissions are separated from each other by means of coarse wavelength division multiplexing (CWDM), where downstream and upstream transmissions take place in the 1.5 μm and 1.3 μm window of the optical fiber, respectively. It is important to note that in power-splitting PONs the bandwidth on the common feeder fiber is shared in both the upstream and downstream directions, giving rise to possible collisions of data packets since the passive splitter/combiner is unable to buffer data packets coming from and going to the different distribution fibers. Clearly, there must be a mechanism in place to allocate bandwidth either statically or dynamically in order to avoid collisions. In the downstream direction, this can be easily done by deploying an appropriate multiplexing technique at the CO, e.g., time division multiplexing (TDM). In the upstream direction, a PON represents a multipoint-to-point multi-access network, in which collisions can be completely avoided by using a suitable medium access control (MAC) protocol. The following four main categories of MAC protocol have been investigated for PONs (Koonen [2006]):

- **Time division multiple access (TDMA)**
  In TDMA, upstream data packets are interleaved in a collision-free manner. This approach requires synchronization of upstream transmissions and the use of burst-mode transceivers at the CO and subscribers.

- **Wavelength division multiple access (WDMA)**
  In a WDMA PON, each subscriber uses a separate wavelength channel to avoid collisions without requiring any synchronization of upstream transmissions on different wavelength channels.

- **Subcarrier multiple access (SCMA)**
  With SCMA, the subscribers modulate their data packet streams on different electrical frequencies, which subsequently modulate the light intensity of their optical transmitters. The optical transmitters at the subscribers may operate at the same nominal wavelength and no synchronization of the different electrical frequency bands is needed.

- **Optical code division multiple access (OCDMA)**
  In an OCDMA PON, each subscriber uses a specific optical code signature to distinguish its encoded packets from those of other subscribers. This approach does not need any synchronization, but requires very high-speed signature sequences and suffers from limited reach due to the increased impact of dispersion.

Of the aforementioned different categories, TDMA PONs have emerged as by far the most important type of currently deployed fiber access networks worldwide, while WDMA PONs are considered a promising candidate for next-generation PONs. Two

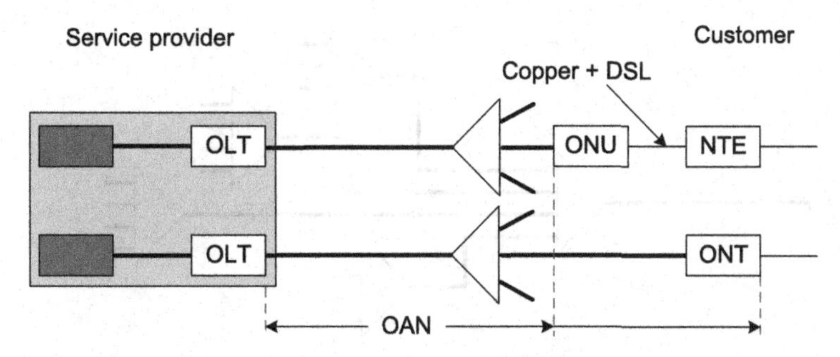

Figure II.1 PON network elements. After Shumate (2008). ©2008 IEEE.

initiatives have led to the standardization of the two most important types of TDMA PONs, known as Gigabit PON (GPON) and Ethernet PON (EPON). GPON emerged from the full service access network (FSAN) alliance, while EPON is an outcome of the Ethernet in the First Mile (EFM) initiative.

The nomenclature introduced by FSAN and shown in Fig. II.1 is now common for most PONs (Shumate [2008]). The service provider's interface to a PON is referred to as *optical line termination* or *optical line terminal (OLT).* The OLT is located at the CO together with other equipment such as switches, servers, and routers (dark boxes in Fig. II.1), which connect the PON to the backbone network. In fiber-to-the-x (FTTx) deployments such as fiber-to-the-cabinet (FTTCab) or fiber-to-the-curb (FTTC), where fiber does not pave all the way to the subscriber, the remote unit is called an *optical network unit (ONU).* The termination of the digital subscriber line (DSL) copper drop line from the ONU to the premises is called the *network termination equipment (NTE).* In the case of fiber-to-the-home (FTTH) or fiber-to-the-building (FTTB) networks, where the fiber terminates at the premises, the active network unit at the subscriber is called an *optical network termination (ONT).* The all-optical passive portion of the PON between OLT and ONU/ONT is referred to as *optical access network (OAN)* or *optical distribution network (ODN).*

Typically, PONs have a tree topology with the OLT located at the root and ONUs/ONTs attached to the leaf nodes of the tree. It is worthwhile to mention that PONs offer some topological flexibility in that the splitter can be placed anywhere in the OAN according to given deployment scenarios (Shumate [2008]). For instance, in rural areas, the splitter may be placed far from the OLT close to a group of connected homes in order to capitalize on the cost sharing of a longer feeder fiber in such a fiber-lean PON configuration. Conversely, in urban areas with high population densities a fiber-rich PON configuration may be preferable by locating the splitter at that OLT and running short point-to-point distribution fibers to the nearby subscribers. Even though such a fiber-rich configuration does not benefit from the cost sharing of a common feeder fiber, there are other benefits that make this configuration attractive in urban areas with generally short distances to the subscribers. Similar to the point-to-point architecture, a fiber-rich PON configuration simplifies upgrades of individual subscribers. Unlike point-to-point solutions, however, it requires only a single optical–electrical interface at the OLT. Another interesting PON configuration can be realized by using multiple

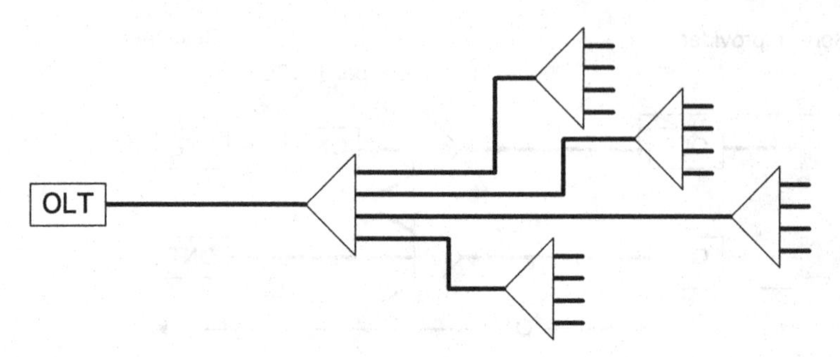

Figure II.2 PON configuration using multiple splitters. After Shumate (2008). ©2008 IEEE.

splitters instead of a single one, as shown in Fig. II.2. This configuration is very helpful to better map splitters to the physical location of home clusters and thereby shorten the length of required distribution fibers to individual homes.

GPON and EPON represent the state of the art of TDMA PONs and are successfully deployed worldwide (Abrams *et al.* [2005]). In the remainder of this part, we elaborate on GPON and EPON and explain them in technically greater detail in Chapter 3 and Chapter 4, respectively. In Chapter 5, we describe recent progress and ongoing work on the design of next-generation PONs, including the above-mentioned WDMA PON.

3 GPON

The gigabit passive optical network (GPON) is an outcome of the full service access network (FSAN) alliance and is specified in the ITU-T G.984.x series of recommendations, which were finalized in February 2004. GPON extends the capabilities of its two predecessors, asynchronous transfer mode (ATM) PON, also known as APON, and broadband PON (BPON). Compared with its predecessors, GPON provides larger splitting ratios, higher up- and downstream data rates, longer reach, improved privacy and security through the use of the Advanced Encryption Standard (AES) algorithm, and a new GPON encapsulation method (GEM) to carry synchronous voice services and data services such as Ethernet in a bandwidth-efficient manner (Shumate [2008]). These extended capabilities of GPON are explained in greater detail in the following.

3.1 Architecture

Figure 3.1 shows the architecture of a GPON network (Effenberger *et al.* [2007]). GPON deploys two different wavelength channels for upstream and downstream communication. The upstream and downstream wavelength channels operate at 1310 nm and 1490 nm, respectively. Several upstream and downstream data rates are specified for GPON, with a maximum data rate of 1.244 Gb/s in the upstream direction and 2.488 Gb/s in the downstream direction. The reach of a GPON network can be as high as 60 km, whereby the differential reach between optical network units (ONUs) must not exceed 20 km. The ITU-T recommendations for GPON allow for a splitting ratio of up to 128. However, practical deployments typically accommodate smaller numbers of ONUs (e.g., 1:64 or 1:32 splitting ratio) due to a limited optical budget of 28 dB.

3.1.1 Video overlay

In addition to the aforementioned pair of upstream and downstream wavelength channels, GPON typically uses another wavelength to realize a video overlay for the distribution of broadcast video signals. The additional video broadcast wavelength channel is combined with the downstream wavelength channel by using a wavelength division multiplexing (WDM) coupler in front of the optical line terminal (OLT), as shown in Fig. 3.1. The video overlay utilizes subcarrier multiplexing (SCM) techniques and can

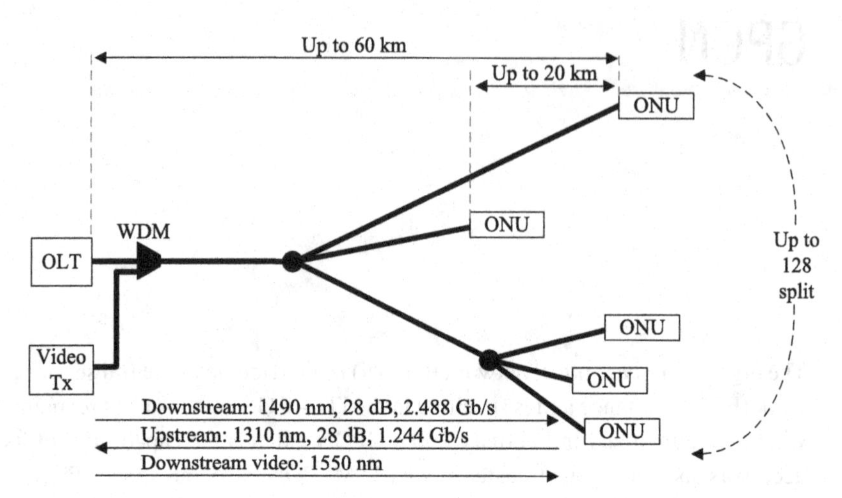

Figure 3.1 GPON architecture. After Effenberger *et al.* (2007). ©2007 IEEE.

deliver either analog video (NTSC, PAL, or SECAM) or digital video (e.g., n-QAM or n-VSB) (Shumate [2008]).

3.1.2 Protection

GPON, and PONs in general, may face network failures such as fiber cuts. Depending on the location of the fiber cut, a single ONU, a subset of ONUs, or all ONUs may be disconnected from the OLT, thus leaving the affected subscribers without service. The following four types of protection were standardized in ITU-T recommendation G.983.1 and may be used to improve the survivability of GPON against link and node failures (Koonen [2006]):

- **Type A**
  Type A protection provides limited protection of the feeder fiber only, whereby a secondary fiber is deployed between the OLT and remote node. After detecting a failure in the primary feeder fiber, optical switches at both ends of the feeder fiber are tossed to switch traffic over to the secondary fiber. Note that re-ranging of the ONUs has to be done after the switch-over is complete.

- **Type B**
  Type B protection involves the duplication of not only the feeder fiber but also the OLT. The backup OLT is on cold standby and is activated when the primary OLT fails.

- **Type C**
  This type of protection implies full GPON duplication, where two separate GPONs are operated in parallel. That is, OLT, feeder fiber, and in addition, each ONU and distribution fiber are duplicated. Unlike type B protection, all secondary devices are hot-standby units, i.e., OLT, ONUs, and optical switches are active during normal

network operation, thus allowing fast switch-over times within 50 ms from the primary equipment to the secondary one.

- **Type D**
  Similar to type C, type D protection backs up the OLT and all ONUs with secondary units. Unlike type C, however, the feeder part and distribution part of GPON are duplicated independently from each other and fast restoration times cannot be provided in type D protection.

For practical deployment, protection types B and C are the most attractive solutions to provide an acceptable trade-off between achievable survivability and required redundancy.

3.2 Wavelength allocation

The majority of GPON and other PON deployments use an additional downstream wavelength channel for video distribution according to the wavelength allocation specified in the ITU-T recommendation G.983.3. Figure 3.2 depicts the wavelength allocation in G.983.3, consisting of the following three wavebands:

- **Upstream band**: The upstream band spans the wavelengths between 1260 nm and 1360 nm. It includes the single upstream wavelength channel, which is typically centered at 1310 nm.
- **Downstream band**: The downstream band is limited from 1480 nm to 1500 nm and comprises the single downstream wavelength channel centered at 1490 nm.
- **Enhancement band**: In addition to the upstream and downstream wavebands, ITU-T G.983.3 specifies a so-called *enhancement band* from 1539 nm to 1565 nm, which is compatible with readily available C-band optical amplifiers, e.g., erbium doped fiber amplifier (EDFA).

The enhancement band can be used to enable additional services. In many deployed GPON networks, the enhancement band is used to realize the above-mentioned video overlay centered at 1550 nm, as shown in Fig. 3.2. The video overlay is broadcast from the OLT to all subscribers. In doing so, there are two different downstream wavelength

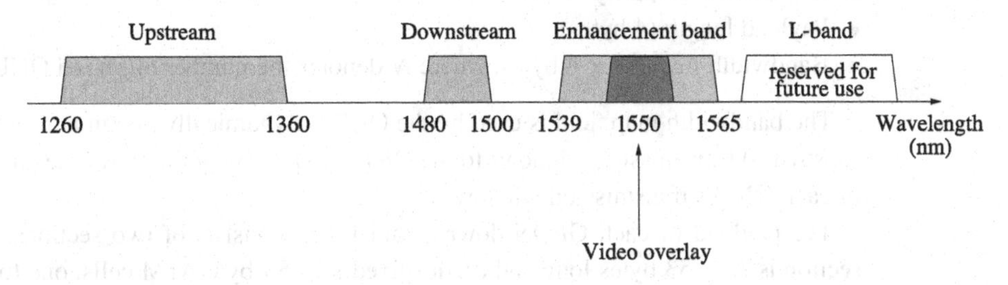

Figure 3.2 ITU-T G.983.3 wavelength allocation.

channels, one wavelength for video at 1550 nm and another wavelength at 1490 nm. The latter might be used for voice and data traffic in order to provide triple-play service offerings, i.e., voice, video, and data. Note that the video overlay requires each ONU to be equipped with a coarse WDM (CWDM) demultiplexer that separates the two downstream wavelengths and an additional receiver operating at 1550 nm.

Figure 3.2 also depicts the L-band, which lies above the enhancement band and covers the wavelength range between 1570 nm and 1610 nm. The L-band contains several wavelength channels that are reserved for future applications. An interesting application for PONs is optical time domain reflectometry (OTDR), which may be run at 1600 nm. OTDR enables testing and troubleshooting and is typically used to locate fiber breaks by detecting the amount of backscattering. In GPON, OTDR can be used to detect a fiber cut in the feeder network between central office and the optical splitter at the remote node. However, OTDR is unable to locate fiber cuts in the distribution network since the OTDR signal is broadcast on all distribution fiber links, whereby each of them may be the source of backscattering.

3.3 GPON encapsulation method

GPON uses periodically recurring frames with a fixed duration of 125 μs for data transfer in both downstream and upstream directions. This periodicity allows GPON to efficiently carry control overhead in the header of each frame and provide quality-of-service (QoS) to delay-sensitive traffic, e.g., voice. In the following, we describe the different downstream and upstream frame formats in greater detail, followed by a description of the GPON encapsulation method (GEM) (Hajduczenia *et al.* [2006]).

3.3.1 Frame formats

Each GPON downstream frame contains a physical control block downstream (PCBd) as header and a payload. The PCBd is broadcast to all ONUs and contains the following fields:

- Physical synchronization (4 bytes)
- Identification (4 bytes)
- Physical layer operation, administration, and management (13 bytes)
- Bit-interleaved parity (1 byte)
- Payload length (4 bytes)
- Bandwidth map ($N \times 8$ bytes, where N denotes the number of served ONUs)

The bandwidth map field is used by the OLT to dynamically assign non-overlapping (upstream) transmission windows to the ONU by specifying the start time and duration of each ONU's transmission window.

The payload of each GPON downstream frame consists of two sections. The first section is $N \times 53$ bytes long and carries fixed-size 53-byte ATM cells, one for each of the N ONUs. The second section is used to transport time division multiplexing (TDM) traffic, e.g., voice, and variable-size data packets, e.g., IP or Ethernet packets, by means of GEM. Note that a GPON network may operate in an ATM-only mode, IP/Ethernet-only mode, or in a mixed mode.

In the upstream direction, each GPON frame contains the upstream transmissions of one or more ONUs according to the schedule given in the aforementioned bandwidth map. During its assigned upstream transmission window, a given ONU is allowed to send backlogged data along with the following preceding control fields:

- Physical layer overhead (variable length with configurable preamble and delimiter)
- Physical layer operation, administration, and management (13 bytes)
- Power leveling sequence (120 bytes)
- Dynamic bandwidth report (variable length)

Note that the physical layer operation, administration, and management field, power leveling sequence field, and dynamic bandwidth report field are optional and are used only upon OLT request. The dynamic bandwidth report field is used by an ONU to report its instantaneous bandwidth demands to the OLT. Its length depends on the number of so-called *traffic containers (T-CONTs)* assigned to the ONU, which are explained in more detail in Section 3.4.

3.3.2 GEM

The GPON encapsulation method (GEM) is a modified version of the ITU-T G.7041 generic framing procedure (GFM), which provides mapping mechanisms to transport different packetized traffic formats over synchronous networks in a bandwidth-efficient manner. GEM enables GPON networks to carry TDM traffic, e.g., voice, and/or variable-size data packet traffic, e.g., Ethernet packets. An individual Ethernet packet contains, apart from its payload, a number of additional control fields, namely, medium access control destination address (MAC DA), MAC source address (MAC SA), type/length, and frame check sequence (FCS). It is important to note that GEM allows for the fragmentation of Ethernet packets in order to increase bandwidth utilization. As a consequence, a client Ethernet packet may be split into fragments spanning multiple GEM payload fields.

As shown in Fig. 3.3, the GEM payload field is preceded by the GEM header, which contains the following fields (Koonen [2006], Hajduczenia *et al.* [2006]):

- Payload length indicator (PLI): contains the length of the payload following the header.
- Port identifier (Port-ID): provides unique traffic identifiers for traffic multiplexing.
- Payload type indicator (PTI): defines the type of carried payload.
- Header error correction (HEC): protects the header against bit errors.

Note that the total length of the GEM header is only 5 bytes, which is considerably smaller than the encapsulation overhead of Ethernet PON.

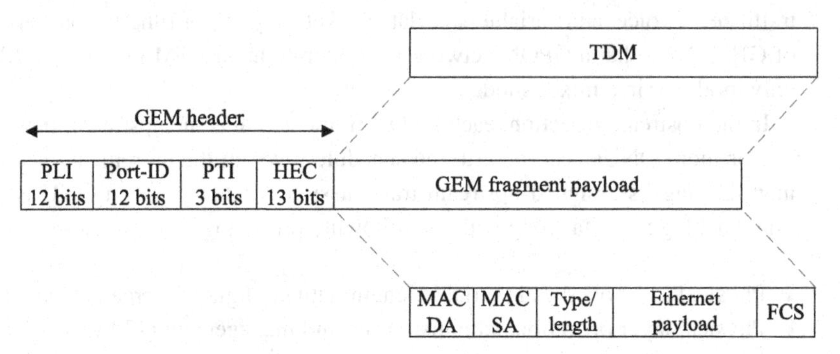

Figure 3.3 GPON encapsulation method (GEM). After Koonen (2006). ©2006 IEEE.

3.4 Bandwidth allocation

Apart from GEM, the GPON transmission convergence layer provides several control functions, including registration and ranging of ONUs, monitoring their status and performance, forward error correction (FEC), encryption, and bandwidth allocation. The transmission convergence layer uses GEM as a connection-oriented encapsulation method. The virtual connection unit of a GPON network is called a GEM port. GEM ports contain traffic flows to and from physical or logical ports of an ONU, whereby each physical/logical port is uniquely identified by its port-ID (see Fig. 3.3). GEM ports are bundled onto one or more T-CONTs. The OLT may assign a single T-CONT or multiple T-CONTs to each ONU. In the latter case, a separate T-CONT may be used for different traffic classes, thus allowing for service differentiation. To facilitate *dynamic bandwidth allocation (DBA)* in GPON, ONUs may use the above-mentioned dynamic bandwidth report field in the GPON upstream frame in order to report their current bandwidth requirements (i.e., queue occupancies) to the OLT. After receiving the bandwidth reports, the OLT informs the ONUs about their assigned upstream transmission windows by using the aforementioned bandwidth map field in the GPON downstream frame. The following two DBA methods are defined for GPON: (*i*) *status-reporting* DBA, which is based on ONU reports using the dynamic bandwidth report field, and (*ii*) *traffic-monitoring* DBA, where the OLT monitors the utilization of assigned T-CONTs and adjusts them according to given traffic loads (Effenberger *et al.* [2007]).

There has been some work on the design and performance evaluation of status-reporting DBA algorithms with the objective to provide QoS support and service differentiation in a bandwidth-efficient manner (Chang *et al.* [2006], Jiang *et al.* [2006]). However, these efforts are negligible compared with the huge number and variety of DBA algorithms proposed for Ethernet PON (EPON) networks, which are described in the next chapter.

4 EPON

Ethernet passive optical network (EPON) has gained a great amount of interest both in industry and academia as a cost-effective solution for broadband access networks, as illustrated by the formation of several forums and working groups, including the EPON forum and the Ethernet in the First Mile (EFM) alliance. EPON carries data encapsulated in Ethernet frames, which makes it easy to carry IP packets and eases the interoperability with installed Ethernet local area networks (LANs). EPON represents the convergence of low-cost Ethernet equipment [switches, network interface cards (NICs)] and low-cost fiber architectures. Furthermore, given the fact that more than 90% of today's data traffic originates from and terminates in Ethernet LANs, EPON appears to be a natural candidate for future first-mile solutions.

The main standardization body behind EPON is the IEEE 802.3ah task force. This task force developed the so-called multipoint control protocol (MPCP) which arbitrates the channel access among central office (CO) and subscribers. MPCP is used for dynamically assigning the upstream bandwidth (subscriber to service provider), which is the key challenge in the access protocol design for EPON. Note that MPCP does not specify any particular dynamic bandwidth allocation (DBA) algorithm. Instead, it is intended to facilitate the implementation of DBA algorithms.

To understand the importance of dynamic bandwidth allocation in EPON, note that the traffic on the individual links in the access network is quite bursty. This is in contrast to metropolitan area networks (MANs) or wide area networks (WANs), where the bandwidth requirements are relatively smooth due to the aggregation of many traffic sources. In an access network, each link represents a single or small set of subscribers that has very bursty traffic conditions, due to a small number of ON/OFF traffic sources. Because of this bursty nature the bandwidth requirements vary widely with time. Therefore, the static allocation of bandwidth to individual subscribers (or sets of subscribers) in an EPON is very inefficient. Employing a DBA algorithm that adapts to instantaneous bandwidth requirements is much more efficient by capitalizing on the benefits of statistical multiplexing. Hence, dynamic bandwidth allocation is a critical feature for EPON design (McGarry *et al.* [2004], Kramer [2005]).

In the following sections, we first describe the EPON architecture and highlight the major operational functions of MPCP. We then provide an overview of state-of-the-art DBA algorithms proposed for EPON. Finally, we briefly describe the salient features of the recently approved IEEE standard 802.3av for 10G-EPON.

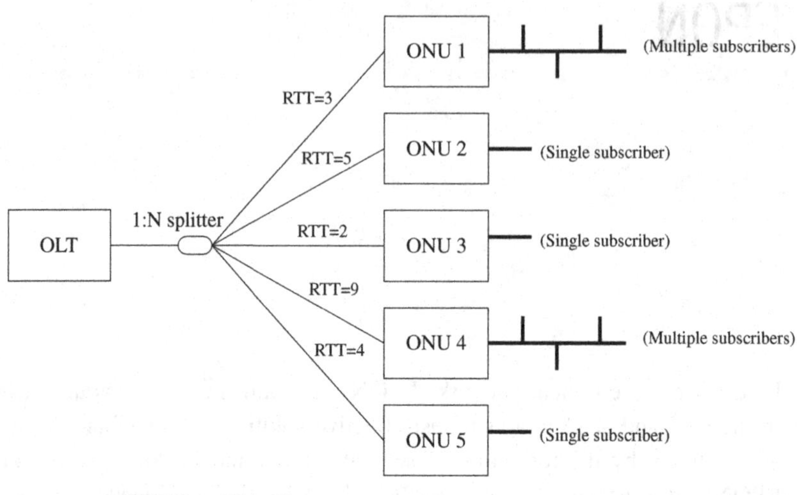

Figure 4.1 EPON architecture.

4.1 Architecture

Typically, an EPON has a physical tree topology with the CO located at the root and the subscribers connected to the leaf nodes of the tree, as illustrated in Fig. 4.1. At the root of the tree is an optical line terminal (OLT) which is the service provider equipment residing at the CO. The EPON connects the OLT to multiple optical network units (ONUs) through a 1:N optical splitter/combiner. An ONU can serve a single residential or business subscriber, or multiple subscribers. Each ONU buffers data received from the attached subscriber(s). In general, the round-trip time (RTT) between OLT and each ONU is different. For instance, in Fig. 4.1 the OLT is connected to five ONUs, each with a different RTT. Due to the directional properties of the optical splitter/combiner the OLT is able to broadcast data to all ONUs in the downstream direction. In the upstream direction, however, ONUs cannot communicate directly with one another. Instead, each ONU is able to send data only to the OLT. Thus, in the downstream direction an EPON may be viewed as a point-to-multipoint network and in the upstream direction, an EPON may be viewed as a multipoint-to-point network (Kramer *et al.* [2003]). Due to this fact, the original Ethernet medium access control (MAC) protocol does not operate properly since it relies on a broadcast medium. Instead, the MPCP protocol is deployed, as discussed in the next section.

An EPON deploys two different wavelength channels, one at 1.5 μm for downstream transmission and one at 1.3 μm for upstream transmissions. Each wavelength channel provides a data rate of 1 Gb/s. In the upstream direction, all ONUs share the upstream wavelength channel. To avoid collisions, several approaches can be used. Wavelength division multiplexing (WDM) is currently considered cost prohibitive since the OLT would require a tunable receiver or a receiver array to receive data on multiple wavelength channels and each ONU would need to be equipped with a wavelength-specific transceiver. At present, time division multiplexing (TDM) is considered a more cost-effective solution. With TDM a single transceiver is required at the OLT and there is just one type of ONU equipment (Kramer *et al.* [2001]). Note that this does not prevent EPON from being upgraded to multiple wavelength channels (WDM) in the future. Given the aforementioned different connectivity in upstream and downstream direction of EPON, the OLT appears to be the best suited node to arbitrate the time sharing of the channel, as discussed next.

4.2 Multipoint control protocol

To increase the upstream bandwidth utilization, the OLT dynamically allocates a variable time slot to each ONU based on the instantaneous bandwidth demands of the ONUs best by means of *polling* (Zheng and Mouftah [2005]). To facilitate DBA and arbitrating the upstream transmissions of multiple ONUs the so-called *multipoint control protocol (MPCP)*, specified in IEEE standard 802.3ah, is deployed in EPON. Beside autodiscovery, registration, and ranging (RTT computation) operations for newly added ONUs, MPCP provides the signaling infrastructure (control plane) for coordinating the data transmissions from the ONUs to the OLT.

As shown in Fig. 4.2, MPCP uses two types of messages to facilitate arbitration: REPORT and GATE. Each ONU has a set of queues, possibly prioritized, holding Ethernet frames ready for upstream transmission to the OLT. The REPORT message is used by an ONU to report bandwidth requirements (typically in the form of queue occupancies) to the OLT. A REPORT message can support the reporting of up to eight queue occupancies of the corresponding ONU. Upon receiving a REPORT message, the OLT passes it to the DBA algorithm module. The DBA module calculates the upstream transmission schedule of all ONUs such that channel collisions are avoided. Scheduling

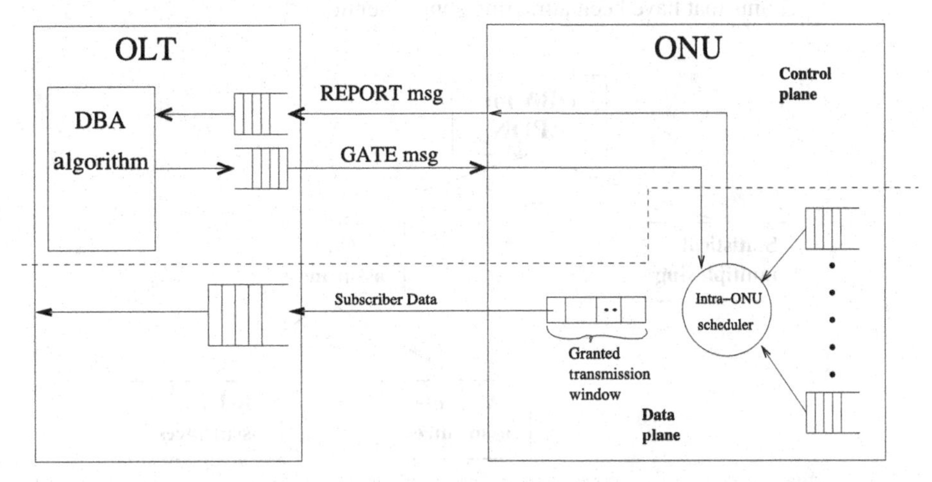

Figure 4.2 Operation of multipoint control protocol (MPCP). After McGarry *et al.* (2004). ©2004 IEEE.

can be done in two ways: inter-ONU scheduling and intra-ONU scheduling. Inter-ONU scheduling arbitrates the transmissions of different ONUs while intra-ONU scheduling arbitrates the transmissions of different priority queues in each ONU. There are two possible implementations. Either inter-ONU scheduling is implemented at the OLT and each ONU performs its own intra-ONU scheduling or both inter-ONU scheduling and intra-ONU scheduling are implemented at the OLT. After executing the DBA algorithm, the OLT transmits GATE messages to issue transmission grants. Each GATE message can support up to four transmission grants. Each transmission grant contains the transmission start time and transmission length of the corresponding ONU. Each ONU updates its local clock using the timestamp contained in each received transmission grant. Thus, each ONU is able to acquire and maintain global synchronization. Each ONU sends backlogged Ethernet frames during its granted transmission window according to the corresponding intra-ONU scheduling. The transmission window may comprise multiple Ethernet frames; packet fragmentation is not allowed. As a consequence, if the next frame does not fit into the current transmission window it has to be deferred to the next granted transmission window.

Note that MPCP does not specify any particular DBA algorithm. MPCP simply provides a framework for the implementation of various DBA algorithms, which are described in greater detail next.

4.3 Dynamic bandwidth allocation (DBA)

According to McGarry *et al.* (2004), DBA algorithms for EPON can be classified into algorithms with statistical multiplexing and algorithms with quality-of-service (QoS) assurances. The latter are further subdivided into algorithms with absolute and relative QoS assurances, as shown in Fig. 4.3. In the following, we discuss the DBA algorithms of each class in greater detail. Finally, we briefly touch on *decentralized* DBA algorithms that have been attracting some attention.

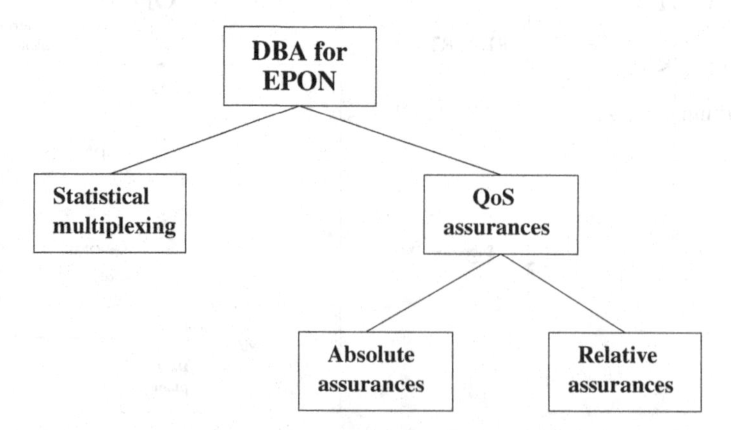

Figure 4.3 Classification of dynamic bandwidth allocation (DBA) algorithms for EPON. After McGarry *et al.* (2004). ©2004 IEEE.

4.3.1 Statistical multiplexing methods

Interleaved polling with adaptive cycle time (IPACT)

In the interleaved polling with adaptive cycle time (IPACT) approach, the OLT polls the ONUs individually and issues transmission grants to them in a round-robin fashion (Kramer *et al.* [2002a]). The grant window size of each ONU's first grant, $G(1)$, is set to some arbitrary value. After n cycles, the backlog (in bytes) in each ONU's transmission buffer, $Q(n)$ (reported queue size), is piggybacked to the current data transmission from the corresponding ONU to the OLT during its grant window $G(n)$. The backlog $Q(n)$ is measured at the instant when the ONU generates the request message, which is piggybacked to the data transmission in cycle n. This backlog $Q(n)$ is used to determine the grant window size of the next grant $G(n + 1)$ of the ONU. In doing so, bandwidth is dynamically assigned to ONUs according to their queue occupancies. If a given ONU's queue is empty, the OLT still grants a transmission window of zero byte to that ONU such that the ONU is able to report its queue occupancy for the next grant. IPACT deploys in-band signaling of bandwidth requests by using escape characters within Ethernet frames instead of sacrificing an entire Ethernet frame for control (as in MPCP), resulting in a reduced signaling overhead. The OLT keeps track of the round-trip times of all ONUs. As a result, the OLT can send out a grant to the next ONU in order to achieve a very tight guard band between consecutive upstream transmissions, resulting in an improved bandwidth utilization. The guard band between two consecutive upstream transmissions is needed to compensate for round-trip time fluctuations and to give the OLT enough time to adjust its receiver to the transmission power level of the next ONU.

In IPACT, each ONU is served once per round-robin polling cycle. The cycle length is not static but adapts to the instantaneous bandwidth requirements of the ONUs. By using a maximum transmission window (MTW), ONUs with high traffic volumes are prevented from monopolizing the bandwidth. The OLT allocates the upstream bandwidth to ONUs in one of the following ways:

- **Fixed service:** This DBA algorithm ignores the requested window size and always grants the MTW size. As a result, the cycle time is constant.
- **Limited service:** This DBA algorithm grants the requested number of bytes, but no more than the MTW.
- **Credit service:** This DBA algorithm grants the requested window plus either a constant credit or a credit that is proportional to the requested window.
- **Elastic service:** This DBA algorithm attempts to overcome the limitation of assigning at most one fixed MTW to an ONU in a round. The maximum window granted to an ONU is such that the accumulated size of the last N grants does not exceed N MTWs, where N denotes the number of ONUs. Thus, if only one ONU is backlogged, it may get a grant of up to N MTWs.

The simulation results reported in (Kramer *et al.* [2002a]) indicate that both the average packet delays and the average queue lengths with the IPACT method with the limited, credit, or elastic service DBA were almost two orders of magnitude smaller compared with the fixed service DBA (fixed service is a static bandwidth allocation) under light traffic loads. Under heavy loads, the average packet delays and average queue lengths for all four types of service were similar. Generally, limited, credit, and elastic service DBA all provided very similar average packet delays and average queue lengths.

In summary, IPACT improves the channel utilization efficiency by reducing the overhead arising from walk times (propagation delay) in a polling system. This is achieved by overlapping multiple polling requests in time. As opposed to static TDM systems, IPACT allows for statistical multiplexing and dynamically allocates upstream bandwidth according to the traffic demands of the ONUs within adaptive polling cycles. Furthermore, IPACT deploys an efficient in-band signaling approach that avoids using extra Ethernet frames for control. By using a maximum transmission window, throughput fairness among the ONUs is achieved. On the downside, this original design for IPACT does not support QoS assurances or service differentiation by means of reservation or prioritization of bandwidth assignment. An IPACT extension to support multiple service classes was developed in (Kramer *et al.* [2002b]), which we discuss in Section 4.3.3.

Control theoretic extension of IPACT

In IPACT, the ONU requests (reports) the amount of backlogged traffic $Q(n)$ as grant for the next cycle. One drawback of this approach is that the request does not take into consideration the amount of traffic arriving at the ONU between the generation of the request message in cycle n and the arrival of the grant $G(n+1)$ for the next cycle at the ONU. As a consequence, the traffic arriving after the generation of a request message is only taken into consideration in the next request message and hence experiences typically a queueing delay of one cycle in the ONU.

To overcome this queueing delay, a control theoretic extension to IPACT was proposed in (Byun *et al.* [2003]). In this extension the amount of traffic arriving at the ONU between two successive requests is estimated and this estimate is incorporated into the grant to the ONU. More specifically, the estimation works as follows. Recall that $Q(n-1)$ denotes the amount of backlogged traffic in the ONU at the instant when the request of cycle $n-1$, which is used by the OLT to calculate the grant $G(n)$, is generated. Let $A(n-1)$ denote the amount of traffic arriving at the ONU between generating the request for cycle $n-1$ and receiving the grant for cycle n. With these definitions, the difference between the grant for cycle n and the amount of traffic backlogged in the ONU when the grant arrives is approximately $D(n) = G(n) - [Q(n-1) + A(n-1)]$. The OLT allocates bandwidth based on the size of the previous grant and the scaled version of the difference reported by the ONUs. More specifically, the grant for cycle $n+1$ is calculated as $G(n+1) = G(n) - \alpha \cdot D(n)$, where α is the gain factor. Using control theoretic arguments it is shown in (Byun *et al.* [2003]) that for piecewise constant traffic with infrequent jumps the system is asymptotically stable for $0 < \alpha < 2$.

Note that this refinement to IPACT essentially views the bandwidth assignment as an automatic control system with the goal to keep the difference $D(n)$ close to zero. A proportional control is proposed for this system with the control gain α. The advantage of this control theoretic approach is that the grant size is typically closer to the size of the backlog at the instant of receiving the grant at the ONU. This in turn results in shorter queueing delays. On the downside, the control system may require careful tuning to achieve a prompt response to changes in the traffic load without creating oscillations in the system. This may be a challenging problem if the traffic load is highly variable.

4.3.2 Absolute QoS assurances

Bandwidth guaranteed polling

The bandwidth guaranteed polling (BGP) method proposed in (Ma *et al.* [2003]) divides ONUs into the two disjoint sets of bandwidth guaranteed ONUs and best-effort ONUs. Bandwidth guaranteed nodes are characterized by their service level agreement (SLA) with the service provider. The SLA specifies the bandwidth this node is to be guaranteed.

The total upstream bandwidth is divided into equivalent bandwidth units, whereby the bandwidth unit is chosen such that the total upstream bandwidth in terms of the bandwidth unit is larger than the number of ONUs. For instance, for a network with 64 ONUs and an upstream bandwidth of 1 Gbps, the equivalent bandwidth unit may be chosen as 10 Mb/s, i.e., the total upstream bandwidth corresponds to 100 bandwidth units. The OLT maintains two tables, one for bandwidth guaranteed ONUs (ONUs with IDs 1 and 4 in the example in Fig. 4.4) and one for best-effort ONUs (ONUs with IDs 2, 3, and 5 in the example in Fig. 4.4). Each table entry (row) has two fields, namely ONU ID and propagation delay from ONU to OLT. The table for bandwidth guaranteed ONUs has as many rows (entries) as there are bandwidth units in the total upstream bandwidth. In the above example, the bandwidth guaranteed ONU table has 100 rows. The table for best-effort nodes is not fixed in size. Entries in the bandwidth guaranteed ONU table are established for each bandwidth guaranteed ONU based on its SLA. If an ONU requires more than one bandwidth unit, then these units are spread evenly through the table as illustrated in Fig. 4.4 for ONU with ID 1, which is guaranteed a bandwidth of two bandwidth units, i.e., 20 Mb/s in the example. Rows in the guaranteed bandwidth ONU table that are not occupied can be dynamically assigned to best-effort nodes. The OLT polls the best-effort ONUs during the rows that are not used by the bandwidth guaranteed ONUs in the order they are listed in the best-effort table.

The OLT begins polling ONUs using the information in the two tables. The OLT polls an ONU by sending a Grant message to grant a window of size G, which is initially set to one bandwidth unit. The ONU decides based on the size of its output buffer if it has enough data to fully utilize the granted transmission window. The ONU sends a reply to the OLT with the amount of the window it intends to utilize B and then transmits this amount of data. The OLT upon receiving a reply from an ONU, checks the amount of the granted window the currently polled ONU intends to use. If B is zero, the OLT immediately polls the next ONU in the table. (Note that this wastes the bandwidth during the round-trip time to that next ONU, whereas the polling to the first ONU can be interleaved with the preceding data transmission to avoid wasting bandwidth.) If B is between zero and some threshold G_{reuse}, whereby $G - G_{reuse}$ specifies the minimum

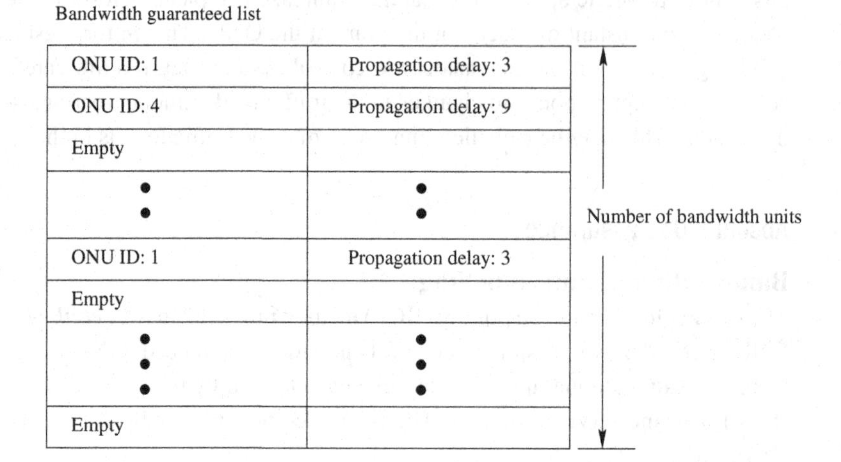

Non-bandwidth guaranteed list

current entry →

ONU ID: 2	Propagation delay: 5
ONU ID: 3	Propagation delay: 2
ONU ID: 5	Propagation delay: 3
⋮	⋮
ONU ID: X	Propagation delay: X

Number of units not fixed

Figure 4.4 Bandwidth guaranteed polling (BGP) tables. After McGarry *et al.* (2004). ©2004 IEEE.

portion of the bandwidth unit that can be effectively shared, the OLT polls the next best-effort ONU ready for transmission and grants it a transmission window $G - B$. Lastly, if B is larger than the threshold G_{reuse}, the OLT will not poll the next ONU until the current grant has passed.

The simulation results reported in (Ma *et al.* [2003]) indicate that for bandwidth guaranteed ONUs with four or more entries, the delays were an order of magnitude smaller than with IPACT. However, for bandwidth guaranteed ONUs with only one entry as well as non-bandwidth guaranteed ONUs, the delays were orders of magnitude larger than with IPACT under light loads and almost an order of magnitude larger than with IPACT under heavy loads. On the other hand, for bandwidth guaranteed ONUs with four or more entries, the queue lengths were similar to IPACT for light loads and were orders of magnitude shorter than with IPACT under heavy loads. However, for bandwidth guaranteed ONUs with only one entry as well as non-bandwidth guaranteed ONUs the queue lengths were orders of magnitude larger than with IPACT under light loads and similar to IPACT under heavy loads. It was also found that the throughput with BGP tends to be lower than the throughput with IPACT, especially at heavy loads.

Overall, the advantage of the bandwidth guaranteed polling approach is that it ensures that an ONU receives the bandwidth specified by its SLA and that the spacing between transmission grants corresponding to SLAs has a fixed bound. The approach also allows for the statistical multiplexing of traffiç into unreserved bandwidth units as well as unused portions of a guaranteed bandwidth unit (i.e., if an ONU does not have enough traffic to use all the bandwidth specified in its SLA). One drawback of the table-driven upstream transmission grants of fixed bandwidth units is that the upstream transmission tends to become fragmented, with each fragment requiring a guard band, which tends to reduce the throughput and bandwidth utilization.

Deterministic effective bandwidth

In (Zhang *et al.* [2003]), a system in which ONUs and OLT employ deterministic effective bandwidth (DEB) admission control and resource allocation in conjunction with generalized processor sharing (GPS) scheduling is developed. In this system, a given ONU maintains several queues, typically one for each traffic source or each class of traffic sources. A given queue is categorized as either a QoS queue or a best-effort queue, depending on the requirements of the corresponding traffic source (class). A given traffic source feeding into a QoS queue is characterized by leaky bucket parameters. The leaky bucket parameters are traffic descriptors widely used in QoS networking and give the peak rate of the source, the maximum burst that the source can send at the peak rate, as well as the long run average rate of the source. A source also specifies the maximum delay it can tolerate. The leaky bucket traffic characterization together with the delay limit of the source (class) are used to determine whether the system can support the traffic in the QoS queues at all ONUs without violating delay bounds (and also without dropping any traffic at a QoS queue) using techniques derived from the general theory of deterministic effective bandwidth.

During the operation of the network, the OLT assigns grants to a given ONU based on the aggregate effective bandwidth of the traffic of the QoS queues at the ONU. Roughly speaking, a given ONU is assigned grants proportional to the ratio of the aggregate effective bandwidth of the traffic of the ONU to the total aggregate effective bandwidth of the traffic of all ONUs supported by the OLT. In turn, a given ONU uses the grants that it receives to serve its QoS queues in proportion to the ratio of the effective bandwidth of the traffic of a queue to the aggregate effective bandwidth of the traffic of the QoS queues supported by the ONU. A given ONU uses the grants not utilized by QoS queues to transmit from best-effort queues.

The advantage of the deterministic effective bandwidth approach is that individual flows (or classes of flows) are provided with deterministic QoS guarantees, ensuring lossless, bounded-delay service. In addition, best-effort traffic flows can utilize bandwidth not needed by QoS traffic flows. One main drawback of the DEB approach is that it requires increased complexity to conduct admission control and update proportions of effective bandwidths of ongoing flows. In particular, conducting admission control and allocating grant resources may result in a significant overhead for short-lived traffic flows (or classes of traffic).

4.3.3 Relative QoS assurances

DBA for multimedia

In dynamic bandwidth allocation for multimedia (Choi and Huh [2002]), traffic in each ONU is placed into one of three priority queues (high, medium, or low). These priorities are then used by the DBA algorithm to assign bandwidth. The sizes of the three priority queues in each ONU are reported to the OLT. The OLT based on the priority queue sizes issues grants separately for each of the priorities in each of the ONUs. In particular, bandwidth is first handed out to the high-priority queues, satisfying all the requests of the high-priority flows. The DBA algorithm then considers the requests from the medium-priority flows. If it can satisfy all of the medium-priority requests with what is left over from the high-priority requests it does so. Otherwise it divides the remaining bandwidth between all medium-priority flows, where the fraction of the bandwidth granted to each medium-priority flow is related to the fraction requested by each flow to the total of all medium-priority requests. Finally, if there is any leftover bandwidth after satisfying the high- and medium-priority requests, this leftover bandwidth is distributed between the low-priority flows in a manner identical to the case where all the medium-priority flows requests cannot be fully satisfied.

Note that in the DBA for multimedia approach, bandwidth is essentially allocated using strict priority based on the requirements of each priority traffic class of the entire PON (all the ONUs connected to a single OLT). One feature of this approach is that the OLT controls the scheduling within the ONU. This comes at the expense of reporting the occupancies of the individual priority queues and issuing multiple grants to each ONU per cycle. Also, the OLT has the additional burden of deciding on the scheduling among the queues in the ONU. Note that the strict priority scheduling based on the traffic classes at the PON level may result in the starvation of ONUs that have only low-priority traffic.

DBA for QoS

The DBA for QoS (Assi *et al.* [2003]) is a method of providing per-flow QoS in an EPON using differentiated services. Within each ONU, priority packet queueing and scheduling is employed per the differentiated services framework. This is similar to the DBA for the multimedia approach, but recall that in the DBA for multimedia the priority scheduling was performed at the PON level (all the ONUs connected to a single OLT). In contrast, in DBA for QoS, the priority scheduling is performed at the ONU level.

Before we proceed to DBA for QoS (Assi *et al.* [2003]), we review the IPACT extension to multiple service classes (Kramer *et al.* [2002b]), which may be viewed as a precursor to DBA for QoS. In (Kramer *et al.* [2002b]), a simulation study is conducted of supporting differentiated service to three classes of traffic with strict priority scheduling inside the ONU. The authors noticed an interesting phenomenon that they dubbed "light-load penalty." What they noticed was that under light loading, the lower priority class experienced a significant average packet delay increase, and also the maximum packet delays for the higher priorities exhibited similar behavior. This appears to be caused by the fact that the queue reporting occurs at some time before the strict priority scheduling is performed, thus allowing higher priority traffic arriving after the queue reporting but before the transmission grant to preempt the lower priority traffic that arrived before the queue reporting. It appears this problem is exacerbated under light loading. The authors discuss two methods for dealing with the "light-load penalty." The first method involves scheduling the packets when the REPORT message is transmitted and placing them in a second stage queue. This second stage queue is the queue that will be emptied out first into the timeslot provided through a grant in a GATE message. The second method involves predicting the number of high-priority packets arriving between the queue reporting and the grant window so that the grant window will be large enough to accommodate the newly arriving high-priority packets. This second method inherently lowers the delay experienced by higher priority traffic compared with the two-stage queueing approach.

In DBA for QoS (Assi *et al.* [2003]), the authors incorporate a method similar to the two-stage queueing approach mentioned above. Specifically, in the DBA for QoS method the packet scheduler in the ONU employs priority scheduling only on the packets that arrive before some $t_{request}$, which is the time at which the REPORT message is sent to the OLT. This avoids the problem of having the ONU packet scheduler request bandwidth based on buffer occupancies at time $t_{request}$ and then actually schedule packets at time t_{grant} to fill the granted transmission window. If this mechanism is not employed, lower priority queues can be starved more severely because higher priority traffic arriving between $t_{request}$ and t_{grant} would tend to take away transmission capacity from the lower priority queues. Note that this problem only arises with strict priority scheduling, which schedules lower priority packets only when the higher priority packet queues are empty. With weighted fair queueing (WFQ), which serves the different priority queues in proportion to fixed weights, this problem would not arise.

In DBA for QoS, each ONU is assigned guaranteed bandwidth in proportion to its SLA. More specifically, let B_{total} denote the total upstream bandwidth. Let w_i denote the weighing factor for ONU i. The weighing factors are set in proportion to the SLA of ONU i, such that the weighing factors of all ONUs supported by the OLT sum to one, i.e., $\sum_i w_i = 1$. ONU i is then assigned the guaranteed bandwidth $B_i = B_{total} \cdot w_i$. Note that the sum of all the guaranteed bandwidths equals the total available bandwidth. In other words, the total upstream bandwidth is divided up among the ONUs in proportion to their SLAs.

For every transmission grant cycle, each of the ONUs requests bandwidth corresponding to its total backlog. If the requested bandwidth is smaller than the guaranteed bandwidth, the difference, i.e., the excess bandwidth, is pooled together with the excess bandwidth from all other lightly loaded ONUs (ONUs whose requested bandwidth is less than their guaranteed bandwidth). This pooled excess bandwidth is then distributed to each of the highly loaded ONUs (ONUs whose requested bandwidth is larger than their guaranteed bandwidth) in a manner that weighs the excess assigned in proportion to the size of their request. Note that this proportional scheduling approach is in contrast to the strict priority scheduling of DBA for multimedia, which does not allocate any bandwidth to lower priority traffic classes until the bandwidth demands of all higher priority traffic classes are met.

We note that DBA for QoS allows for the option of sending the individual priority queue occupancies to the OLT via REPORT messages (a REPORT message supports reporting queue sizes of up to eight queues) and having the OLT generate transmission windows for each individual priority queue (the GATE message supports sending up to four transmission grants). This option puts the priority scheduling that would otherwise be handled by the ONU under the control of the OLT.

DBA for QoS (Assi *et al.* [2003]) also considers the option of reporting the queue size using an estimator for the occupancy of the high-priority queue. The estimator makes a one-step prediction of the traffic arriving at the high-priority queue between the time of the report and the time of the grant. In particular, the amount of traffic arriving at the high-priority queue between report and grant in a cycle $n - 1$ is used to estimate the arrival in cycle n. The ONU then reports in cycle n the actual backlog at the time of request plus the estimated new arrivals until the time of the grant.

The simulations reported in (Assi *et al.* [2003]) compare the average and maximum delays for the proposed DBA for QoS scheme for the service classes best effort, assured forwarding, and expedited forwarding with the delays achieved with a static bandwidth allocation to the individual ONUs. It was found that the proposed DBA for the QoS scheme achieves significantly smaller delays, especially at high loads. This is primarily due to the statistical multiplexing between the different ONUs permitted by the DBA for QoS. It was also found that the proposed DBA for QoS scheme is quite effective in differentiating the delays for the different service classes, with the highest priority expedited forwarding class achieving the smallest delays. The simulations in (Assi *et al.* [2003]) also considered the average utilization of the upstream bandwidth and found that the proposed DBA for QoS schemes achieve around 90% utilization compared with around 50% with static bandwidth allocation.

4.3.4 Decentralized DBA algorithms

Note that all the above-mentioned DBA algorithms are centralized schemes. The OLT acts as the central control unit by performing inter-ONU scheduling or both inter-ONU and intra-ONU scheduling. Alternatively, *decentralized* DBA algorithms and distributed scheduling were investigated in (Foh *et al.* [2004], Sherif *et al.* [2004]). To enable distributed scheduling, however, the original EPON architecture has to be modified such that each ONU's upstream transmission is echoed at the splitter to all ONUs, each equipped with an additional receiver to receive the echoed transmissions. In doing so, all ONUs are able to monitor the transmission of every ONU and to arbitrate upstream channel access in a distributed manner, similar to Ethernet LANs. Note that in such alternate EPON solutions both inter-ONU and intra-ONU scheduling take place at the ONUs without the participation of the OLT. The reported performance results show that such decentralized EPONs and DBA algorithms are able to provide high bandwidth utilization. We refer the interested reader to (Foh *et al.* [2004], Sherif *et al.* [2004]) for further details on decentralized DBA algorithms.

4.4 10G-EPON

To address the continuously increasing demand of high-speed data communication in fiber access networks, the 10G-EPON has emerged as a promising candidate for next-generation high data rate access systems by extending the EPON operation to ten-fold data rates of 10 Gb/s (Tanaka *et al.* [2010]). The 10G-EPON standardization was initiated by the IEEE 802.3 working group in March 2006. The IEEE 802.3av task force was formed in July 2006 with the aim of developing the physical layer specification and management parameters. In September 2009, the IEEE 802.3av standard was finalized and the development of the specification was completed. 10G-EPON is expected to provide a tenfold capacity increase at the cost of three times the port price of the currently established Ethernet hardware (Hajduczenia *et al.* [2008]).

The 10G-EPON standard provides symmetric 10 Gb/s downstream and upstream, as well as asymmetric 10 Gb/s downstream and 1 Gb/s upstream data rates. In order to provide backward compatibility with the existing and widely deployed EPON, the OLT in a 10G-EPON is equipped with dual-rate receivers for receiving data from 1G- and 10G-ONUs. Furthermore, the downstream transmission channels are separated for sending downstream data and control traffic to 1G- and 10G-ONUs.

The IEEE 802.3av task force focused only on the physical layer, while maintaining complete backward compatibility with 1 Gb/s EPON equipment. The MAC protocol of EPON (MPCP) remains unchanged. Similar to EPON, the MAC protocol in 10G-EPON operates on the basis of ONUs informing the OLT of their upstream bandwidth requirements, and the OLT scheduling and granting bandwidth to the ONUs to transmit their upstream data.

One major difference between EPON and 10G-EPON is that the latter supports both symmetric 10 Gb/s downstream and upstream, and asymmetric 10 Gb/s downstream and 1 Gb/s upstream data rates, while EPON provides only 1 Gb/s symmetric data rate. The line coding for the 10G-EPON and EPON is also different. The 64B/66B line coding in 10G-EPON reduces the bit-to-baud overhead to 3%, compared with the 25% overhead in EPON which is incurred by 8B/10B line encoding. The burst signal format of 10G-EPON is similar to that of EPON, except that the receiver settling time of 10G-EPON is twice that of EPON, i.e., 800 ns in 10G-EPON and 400 ns in EPON. The laser on/off time and clock data recovery time of both EPON standards are the same (512 ns and 400 ns, respectively) (Tanaka *et al.* [2010]).

The wavebands utilized for upstream (US) and downstream (DS) transmissions of conventional 1G-EPON and recently standardized 10G-EPON networks are illustrated in Fig. 4.5 and Fig. 4.6, respectively. EPON allocates a 100 nm waveband centered at 1310 nm for upstream (US) transmission and a 20 nm window centered at 1490 nm for downstream (DS) transmission. The downstream wavelength of 10G-EPON is allocated in a window between 1575 and 1580 nm (with a typical value of 1577 nm), which is

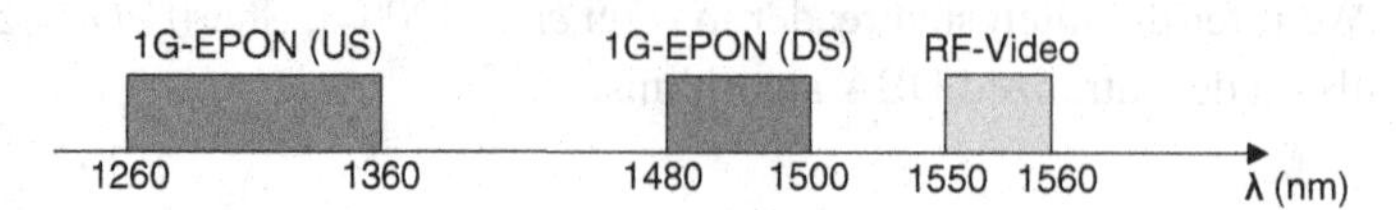

Figure 4.5 Waveband allocation in EPON.

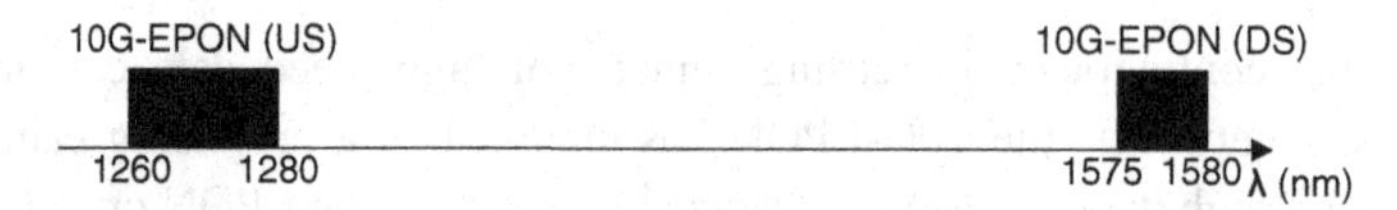

Figure 4.6 Waveband allocation in 10G-EPON.

outside of the analog RF video distribution band. Conversely, the upstream wavelength of 10G-EPON is allocated in a 20 nm window centered at 1270 nm, which is completely covered by a part of the EPON upstream waveband.

The backward compatibility requirement of the new and existing EPONs introduces several technical challenges and difficulties on the specification work such as a high power budget exceeding 30 dB for symmetric 10 Gb/s transmission, conflicts in wavelength allocation, and dual-rate burst-mode operation at the OLT receiver (Hajduczenia *et al.* [2008]). Two main techniques are employed for achieving the coexistence of 10G-EPON with 1G-EPON (and analog RF video distribution) systems: WDM overlay in the downstream direction and a dual-rate burst-mode receiver in the upstream direction to support a dual-speed TDM. In the downstream direction, since the wavelength bands are distinct, a WDM overlay is a straightforward way to provide the coexistence with EPON. On the contrary, the upstream wavelength band of 10G-EPON is in fact a subset of the EPON waveband. Hence, a dual-rate burst-mode operation is the only remaining option to retain the coexistence requirement by using dual-speed TDM (Tanaka *et al.* [2010]). For this purpose, the OLT is equipped with a dual-rate receiver. The received dual-rate signal can be separated at the OLT via either an optical domain or an electrical domain split. The choice of which scheme to use is an implementation matter. In general, the electrical domain split is considered more practical due to its simple configuration (Tanaka *et al.* [2010]). At the MAC layer, the OLT provides three kinds of MAC instances for operating on symmetric and asymmetric data rates; namely, the OLT supports 1/1 Gb/s, 10/1 Gb/s, and 10/10 Gb/s MAC instances.

5 Next-generation PON

GPON and EPON, described at length above in Chapter 3 and Chapter 4, respectively, represent the two most important Gigabit-class passive optical networks (PONs) that are widely deployed in the United States, Europe, and Asia Pacific region. Given the ever increasing bandwidth demand from consumer and business applications, current PONs are expected to evolve into *next-generation PONs (NG-PONs)* over the next couple of years. GPON and EPON are expected to coexist for the foreseeable future as they evolve into NG-PONs. Clearly, one way to realize NG-PONs is to increase the line rate of current Gigabit-class PONs to 10 Gb/s. A good example of this approach is the IEEE 802.3av 10G-EPON standard, which was approved in September 2009 (see Section 4.4). NG-PONs are mainly envisioned to (i) achieve higher performance parameters, e.g., higher bandwidth per subscriber, increased splitting ratio, and extended maximum reach, than current GPON/EPON architectures, and (ii) broaden GPON/EPON functionalities to include, among others, the consolidation of optical access, metro, and backhaul networks, and the support of topologies other than conventional tree structures. Network operators are seeking NG-PON solutions that can transparently coexist with legacy PONs on the existing fiber infrastructure and enable gradual upgrades in order to avoid costly and time-consuming network modifications and stay flexible for further evolution paths.

According to Kani *et al.* (2009), NG-PON technologies can be divided into the following two categories:

- **NG-PON1**
  This type of technology allows for an *evolutionary* growth of existent Gigabit-class PONs and supports their coexistence on the same optical distribution network (ODN). The coexistence is intended to let customers individually be upgraded without incurring any service discontinuity for other customers.

- **NG-PON2**
  This category of enabling technologies envisions a *revolutionary* upgrade of current PONs, giving rise to disruptive NG-PONs without any coexistence requirements with existent Gigabit-class PONs on the same ODN.

Figure 5.1 shows the migration from current state-of-the-art GPON and EPON to near-term NG-PON1 and mid- to long-term NG-PON2 broadband access solutions

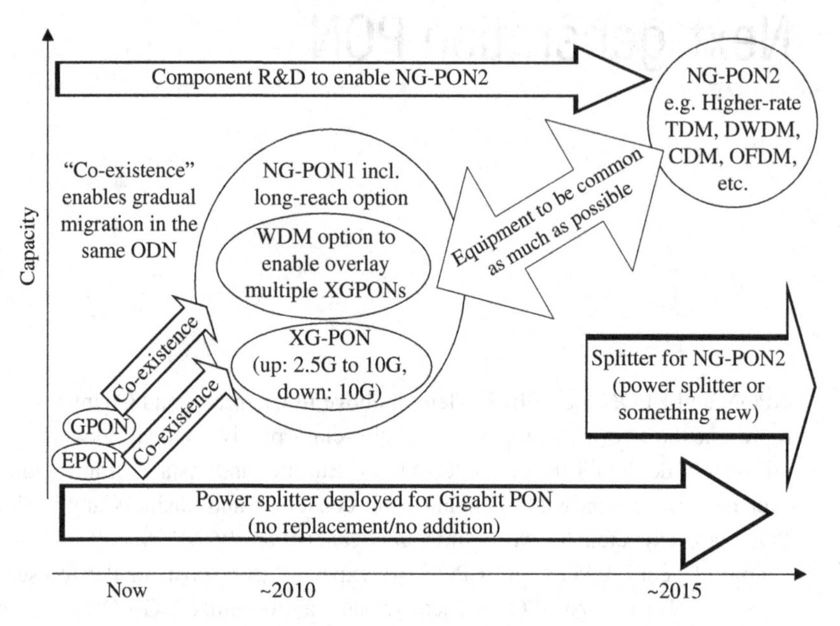

Figure 5.1 Next-generation PON (NG-PON) roadmap. After Kani *et al.* (2009). ©2009 IEEE.

(Kani *et al.* [2009]). NG-PON1 technologies include a number of performance-enhancing options. One of the most promising NG-PON1 solutions is the so-called XG-PON that provides 10 Gb/s upstream and/or downstream (the Roman numeral X stands for ten). More precisely, an asymmetric XG-PON supporting 10 Gb/s downstream and 2.5 Gb/s upstream is called an XG-PON1. Whereas a symmetric XG-PON offering 10 Gb/s in both downstream and upstream directions is referred to as an XG-PON2. Other interesting NG-PON1 technologies include the use of wavelength division multiplexing (WDM) to realize an overlay of multiple XG-PONs on the ODN or the deployment of reach extenders to enable long-reach PONs. Note that in all possible NG-PON1 solutions the ODN is left untouched, keeping the power splitter at the remote node in place. Several NG-PON1 candidates have been proposed by the full service access network (FSAN) alliance, as discussed in Section 5.1.

In the long term, NG-PON1 will be gradually replaced by NG-PON2 solutions after resolving a number of issues related to the research and development (R&D) of advanced optical network components and enabling technologies. NG-PON2 technologies include a wide variety of technical candidates, e.g., higher-rate time division multiplexing (TDM), dense WDM (DWDM), code division multiplexing (CDM), orthogonal frequency division multiplexing (OFDM), and others. We will describe some of the most important NG-PON2 technologies in greater detail below in Section 5.2. It is important to note that future NG-PON2 optical access networks may use a different device at the remote node instead of a simple power splitter, even though for cost reasons NG-PON1 and NG-PON2 solutions should have as much equipment as possible in common (Kani *et al.* [2009]).

Before elaborating on possible NG-PON1 and NG-PON2 candidate technologies in the remainder of this chapter, let us briefly look at some of the key requirements they have to meet (Kani *et al.* [2009]). To reduce the operational expenditures (OPEX) and greenhouse gas emissions, NG-PONs should deploy power saving techniques. Toward this end, XG-PONs should support at least the following two modes: (*i*) full-service mode and (*ii*) sleep mode. In addition, a third mode should be considered that allows ONUs to reduce their use of backup batteries during power outages in order to support only the lifeline service carrying interface (e.g., voice service). Furthermore, as network operators think of deploying NG-PON-based infrastructures such as 3G/4G backhaul, the OLT and ONUs must be able to provide accurate synchronization to the mobile cell sites. Finally, PON supervision techniques without significantly increasing the capital expenditures (CAPEX) and compromising on the bandwidth for services are required to be able to differentiate between faults in the ODN and faults in the electronics connected to it.

5.1 NG-PON1

FSAN considers different candidate architectures for near-term NG-PON1 systems, including XG-PON, long-reach XG-PON, and WDM XG-PON. In the following, we briefly review the salient features of these candidate systems. For a more detailed description of the involved technical challenges and coexistence issues in the time or wavelength domain we refer the interested reader to (Effenberger *et al.* [2009a], Zhang *et al.* [2009], Effenberger *et al.* [2009b]).

5.1.1 XG-PON

As we have seen above, XG-PONs offer data rates of up to 10 Gb/s and come in two flavors: (*i*) asymmetric XG-PON1 and (*ii*) symmetric XG-PON2. One of the most significant challenges of XG-PON is the realization of burst-mode transceivers that are able to operate at higher data rates. Another important challenge is dispersion, which becomes more of a concern at 10 Gb/s and may cause intersymbol interference (ISI). To address these challenges, different ways have been studied to implement XG-PON. One approach might be to divide the 10 Gb/s channel into four different wavelength channels, each operating at 2.5 Gb/s.

5.1.2 Long-reach XG-PON

To extend the reach of XG-PON, the deployment of an advanced transmitter and receiver with optical pre- and post-amplification might be required. By using wavelength-controlled ONUs (e.g., cooled laser sources) the receiver sensitivity and thus optical power budget is improved by multiple dB. However, the cost of wavelength-controlled transmitters poses a challenge in cost-sensitive access networks. Alternatively, optical amplifiers may be placed in the ODN to compensate for the increased propagation and splitting losses of long-reach PONs (Shea and Mitchell [2007]). (Strictly speaking, long-reach PONs that use in-line optical amplifiers are not passive anymore; a more accurate term for such a PON would be *active optical network (AON)*, since optical amplifiers are active network elements that require their own power supply.) Long-reach XG-PONs are able to displace costly electronics and simplify network operation by combining access and metro networks into one.

5.1.3 WDM XG-PON

One promising approach to avoiding the higher cost of wavelength-controlled transmitters is to combine multiple XG-PONs by means of a WDM multiplexer/demultiplexer and use so-called *colorless* (i.e., wavelength-independent) ONUs. In the resultant WDM XG-PON, each XG-PON runs on a different set of wavelength channels and doesn't interfere with other wavelength-multiplexed XG-PONs. Colorless ONUs require either only a broadband light source or no light source at all, resulting in decreased costs, simplified maintenance, and reduced stock inventory issues. Colorless ONUs are not restricted to a specific wavelength channel. They are able to use any of the available wavelength channels for upstream transmission and downstream reception. A promising approach toward realizing low-cost colorless ONUs is the use of a reflective semiconductor optical amplifier (RSOA) at the ONU for bidirectional transmission on any given wavelength channel. The RSOA performs remote modulation, amplification, and reflection of an optical seed signal sent by the OLT. The optical seed signal can be either a modulated signal carrying downstream data or an unmodulated empty carrier. In the former case, the colorless ONU reuses the modulated carrier by means of remodulation techniques, such as frequency shift keying (FSK) for downstream and on–off keying (OOK) for upstream. The feasibility of RSOA-based WDM XG-PON using electronic equalization and forward error correction (FEC) techniques was demonstrated in (Cho *et al.* [2008]) for an optical reach of >20 km. A long-reach WDM XG-PON with an extended length of 100 km and comprising 17 wavelength-multiplexed XG-PONs, each with up to 256 ONUs, was experimentally investigated in (Talli and Townsend [2006]).

5.2 NG-PON2

Recall from above that NG-PON2 systems are considered long-term broadband access solutions that may capitalize on a wide variety of different enabling technologies and may also replace the passive splitter at the remote node of conventional PONs with another device. In the following, we highlight an NG-PON2 architecture that deploys a wavelength-routing device at the remote node. In addition, we elaborate on optical code division multiple access (OCDMA) and orthogonal frequency division multiple access (OFDMA) PONs as possible NG-PON2 candidates.

5.2.1 Wavelength-routing PON

A wavelength-routing PON can be realized by replacing the wavelength-broadcasting power splitter with an *arrayed-waveguide grating (AWG)* (Koonen [2006]). The AWG is a passive athermal (i.e., temperature-insensitive) device that acts as a wavelength demultiplexer in the downstream direction and as a wavelength multiplexer in the upstream direction. In an AWG-based WDM PON, each ONU is connected to the OLT by one or more wavelength channels, which are separated from each other by the periodicity of the AWG's frequency response (also known as the free spectral range (FSR) of the AWG). As a result, each wavelength forms a point-to-point (PtP) connection between a given ONU and the OLT. Each wavelength per FSR is dedicated to a different ONU. Hence, there is no need for any medium access control (MAC) protocol in a wavelength-routing PON. On the downside, statistical multiplexing cannot be applied in a wavelength-routing PON in order to improve the utilization of wavelength channels. Wavelength-routing PONs offer a number of advantages. Among others, service and capacity upgrades can be done for individual ONUs without affecting the remaining ONUs. Wavelength-routing PONs provide improved privacy since each wavelength channel is dedicated to a single ONU. Similarly to point-to-point architectures, wavelength-routing PONs offer the aforementioned benefits of point-to-point connections. Unlike point-to-point architectures, however, wavelength-routing PONs benefit from cost-sharing a common feeder fiber.

Wavelength-routing PONs can be realized for a few wavelengths using low-cost coarse WDM (CWDM) or for a larger number of wavelengths using more expensive DWDM technologies. Despite some issues related to the cost and temperature stability, DWDM devices provide the following important advantages (Shumate [2008]):

- **Capacity**
  Due to the fact that a wavelength-routing PON provides a dedicated wavelength channel to each ONU, any data rate and modulation format can be used now or as a later upgrade.

- **Security**
  Using dedicated upstream and downstream wavelength channels helps eliminate security issues and simplify network operation.

- **Medium access**
  Another benefit of using dedicated wavelength channels is the fact that no MAC protocol is needed.

- **Power budget**
  While in conventional power-splitting PONs with N attached ONUs the splitting loss equals 10 log N dB, the insertion loss of an AWG is only a few dB, thus increasing the power budget by approximately 9–12 dB for 16–32 ONUs.

Many research activities have been carried out to minimize the cost of wavelength-routing PON components and explore new architectures with improved cost-sharing and growth capabilities. For a recent survey on the latest developments of alternative NG-PON2 architectures we refer the interested reader to (Maier [2009]).

5.2.2 OCDMA PON

OCDMA is viewed as a candidate technology for future PON access networks (Fouli and Maier [2007]). An OCDMA PON uses a conventional tree topology with a power splitter located at the remote node. Each ONU contains an encoder and decoder with unique fixed codes. The OLT may contain all encoder–decoder pairs required for communication with each ONU or a smaller number of tunable encoder–decoders.

Most OCDMA PON propositions suffer from a lack of transitional models that take into account legacy systems. Usually, entire OCDMA networking systems are presented as forklift upgrade alternatives. The main barriers to OCDMA PON deployment reside in the physical layer. Many design issues must be demonstrated if OCDMA PONs are to emerge. Beyond OCDMA transceiver operation, research must control noise source and multiple access interference (MAI) to achieve acceptable bit error rate (BER) levels. Most importantly, OCDMA transceivers at the OLT and ONUs must be manufactured and deployed at acceptable costs.

An important step toward commercial viability might be the enabling of gradual migration paths from wavelength and time division multiple access (WDMA/TDMA) PONs to OCDMA PONs. A gradual migration path offers partial implementations that postpone some of the research elements required for full OCDMA PON deployment. Figure 5.2 illustrates a possible migration scenario (Fouli and Maier [2007]). The shaded ONUs and transceivers (TR) use OCDMA. The conventional splitter–combiner at the remote node may be replaced with wavelength multiplexers and waveband selectors (WS). WS devices are passive WDM multiplexers with lower wavelength granularity and are typically used to separate data and signaling wavebands.

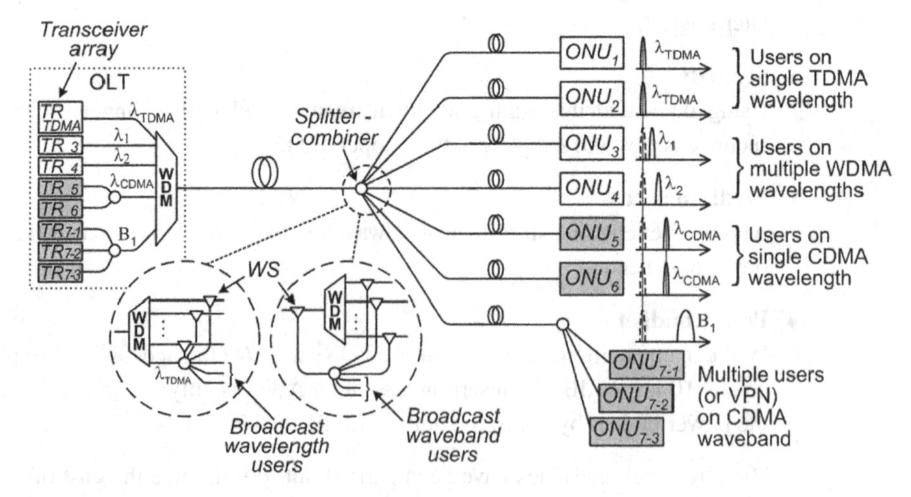

Figure 5.2 OCDMA PON architecture with coexisting legacy TDM and WDM ONUs. After Fouli and Maier (2007). ©2007 IEEE.

To maintain broadcast signals in the wavelength multiplexer configurations, the broadcast wavelengths (TDMA, OCDMA) must be separated from the point-to-point wavelengths (WDMA) so as to split their power among all users, whereby the WS devices separate and merge the broadcast signals. Note that the PON configuration in Fig. 5.2 allows both TDM and WDM ONUs to remain untouched and coexist with the OCDMA upgrade of individual ONUs, including the potential use of the legacy TDMA wavelength channel for broadcast services.

An interesting approach to overcoming the aforementioned barriers of OCDMA deployments is the use of optical coding (OC) technologies for control and/or management plane functionalities rather than data transmission. OC is the process by which a code is inscribed into and extracted from an optical signal. Although a prerequisite for OCDMA, OC boasts a wide range of novel and promising applications. An interesting example is the use of OC technologies for enhanced real-time bandwidth allocation in PONs in order to improve their guaranteed quality-of-service (QoS) performance (Fouli *et al.* [2009]).

5.2.3 OFDMA PON

Recently, the OFDMA PON has been proposed as a possible NG-PON2 technology (Cvijetic *et al.* [2010]). An OFDMA PON may be viewed as a hybrid approach that combines pure OFDMA with TDMA, whereby OFDM subcarriers can be dynamically assigned to different services in different time slots. In doing so, OFDMA PON provides an improved bandwidth flexibility. For downstream transmission of heterogeneous services, the OLT divides the frequency and time domain of an OFDMA frame and informs all ONUs by broadcasting the established time-frequency schedule over non-reserved OFDM subcarriers during pre-configured time slots. For reception of downstream data and transmission of upstream data, each ONU selects its pre-assigned subcarrier(s) according to the aforementioned time-frequency schedule. Note that each ONU may be assigned a single or multiple subcarriers in a given time slot. For cost-efficiency, it is desirable to deploy colorless ONUs rather than wavelength-specific ones. In (Cvijetic *et al.* [2010]), a centralized light source OFDMA PON with colorless ONUs was presented and it was experimentally shown that this architecture is able to achieve 20 Gb/s upstream transmission on a single wavelength channel over a 20 km long ODN with a 1:32 optical split. With respect to downstream transmission in an OFDMA PON, the highest data rate over a single wavelength channel was demonstrated in (Qian *et al.* [2010]), achieving 108 Gb/s per wavelength using polarization multiplexing and direct detection.

Part III

Wireless access networks

6 WiFi

Wireless fidelity (WiFi) has been envisioned by the WiFi alliance as a single worldwide adopted standard for high-speed wireless local area networking. The term WiFi denotes wireless local area network (WLAN) technology based on IEEE 802.11 specifications. In this chapter, we provide an overview of the salient features and most important specifications of legacy and next-generation WLANs.

6.1 Legacy WLAN

WLANs based on IEEE 802.11 have become very popular in providing different data services. Figure 6.1 shows the general WLAN architecture, where an access point (AP) is connected to the Internet and/or other WLANs through a wired network infrastructure, referred to as the distribution system (DS). In this architecture, wireless stations (STAs) communicate with their associated AP using the medium access control (MAC) protocols defined in the IEEE 802.11 specifications.

Due to the use of unlicensed frequency bands (2.4 GHz with 14 distinct channels and 5 MHz) in IEEE 802.11b/g with data rates of up to 11/54 Mbps, WiFi networks have gained much attention (Kuran and Tugcu [2007]). During the last decade, various standards and/or amendments have been approved or initiated to enhance IEEE 802.11 based WiFi technology. Table 6.1 summarizes the IEEE 802.11 WiFi standard family.

The initial IEEE 802.11 physical (PHY) layer includes: (*i*) frequency hopping spread spectrum (FHSS), (*ii*) direct sequence spread spectrum (DSSS), and (*iii*) infrared (IR). IEEE 802.11b uses high-rate DSSS (HR-DSSS), while IEEE 802.11g deploys orthogonal frequency division multiplexing (OFDM). The IEEE 802.11 MAC layer deploys the distributed coordination function (DCF) as a default access technique. In this contention-based scheme, STAs associated with the AP use their air interfaces for sensing the channel availability. If the channel is idle, the source STA sends its data to the destination STA through the associated AP. If more than one STA tries to access the channel simultaneously a collision occurs.

The standard defines the carrier sense multiple access/collision avoidance (CSMA/CA) access protocol to avoid collisions. In CSMA/CA, before initiating a transmission, the source node stores generated frames in its buffers and senses the wireless channel. The DCF protocol employs binary exponential backoff to access the medium. DCF does not use any collision detection function as the nodes cannot detect collisions

Table 6.1. IEEE 802.11 WiFi standard family (Kuran and Tugcu [2007]).

Standard/ Amendment	Description
802.11	Up to 2 Mb/s, 2.4 GHz
802.11a	Up to 54 Mb/s, 5 GHz
802.11ac	Gigabit very high-throughput WLAN, lower than 6 GHz
802.11ad	Gigabit very high-throughput WLAN, 60 GHz
802.11b	Up to 11 Mb/s, 2.4 GHz
802.11d	International roaming extension for 5 GHz band
802.11e	QoS enhancements
802.11g	Up to 54 Mb/s, 2.4 GHz
802.11h	Spectrum managed 802.11a for satellite and radar compatibility
802.11i	Security enhancements
802.11j	Specific extensions for Japan
802.11k	Radio resource measurement extensions
802.11n	Up to 600 Mb/s high-throughput WLAN, 2.4 GHz
802.11p	Wireless access for vehicular environment
802.11r	Fast roaming between WLANs
802.11s	Mesh topology support
802.11u	Interworking between different WLANs
802.11v	Wireless network management
802.11w	Protected management frames
802.11y	3.65–3.7 GHz physical layer standard

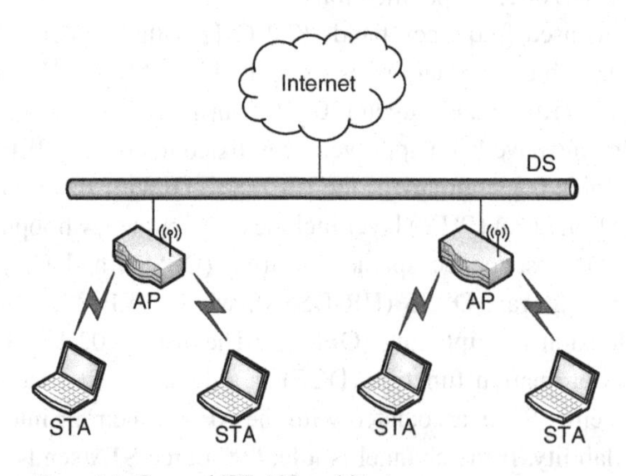

Figure 6.1 General WLAN architecture.

by listening to their own transmissions. Instead, it uses a handshaking method with positive acknowledgment of successfully received frames. When the source node generates a new frame for transmission, it first monitors the channel activity. If the channel is detected idle for a period of time called DCF interframe space (DIFS), the node can transmit immediately. Note that the value of DIFS may vary depending on the deployed standard. For instance, it is set to 50 μs and 34 μs in IEEE 802.11b and IEEE 802.11a/g standards, respectively.

If the channel is busy, the node will wait until the end of the transmission and a random backoff interval is then selected. The backoff counter is decremented as long as the channel is sensed idle, is stopped when channel activity is detected, and is reactivated when the channel is sensed idle for more than a DIFS again. The period following an idle DIFS is slotted and the backoff time counter is measured in terms of slot time. The slot time is the time needed for any node to detect transmissions from the other nodes. It accounts for the propagation delay, the time needed to switch from the receiving to the transmitting state, and the time to notify the MAC layer about the state of the channel. The backoff time is uniformly chosen in the range $(0, CW - 1)$, where CW denotes the current contention window. At the first transmission attempt, CW is set to the minimum contention window CW_{min}. After each unsuccessful transmission, CW is doubled until it reaches the maximum contention window CW_{max}. The node transmits its frame when the backoff counter reaches zero.

It is important to note that DCF applies the automatic response request (ARQ) mechanism to enhance network reliability. Upon reception of a frame, the destination node sends an acknowledgment (ACK) frame to the sender after a short interframe space (SIFS). Similar to DIFS, the value of SIFS is specified in the respective standard. For instance, SIFS is set to 10 μs and 16 μs in IEEE 802.11b and IEEE 802.11a/g, respectively. If the source node does not receive the ACK within a specified Ack timeout or it detects the transmission of a different frame on the channel, it reschedules the frame transmission according to the previous backoff rules. Figure 6.2 illustrates how nodes access the wireless medium using DCF.

The point coordination function (PCF) is another technique that may be used in the MAC layer (Kuran and Tugcu [2007]). In PCF, the data transmission is arbitrated in two modes: (*i*) centralized mode, where the AP polls each STA in a round-robin fashion, and (*ii*) contention-based mode, which works similarly to DCF, but each channel access is controlled and monitored by the AP. In PCF, the AP has a higher channel access priority than STAs by using a shorter interval than DIFS, which is called PCF interframe space (PIFS). Note that a given STA is allowed to send a data packet only after it is polled by the AP, whereby the size of the data packet is restricted by the size of the WLAN MAC service data unit (MSDU),

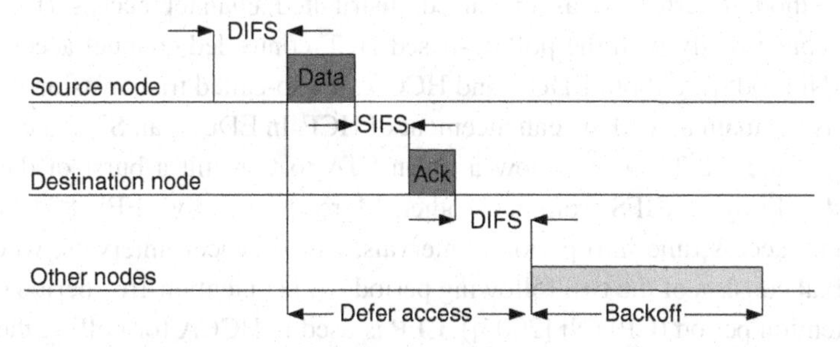

Figure 6.2 Channel access in WLAN networks using DCF.

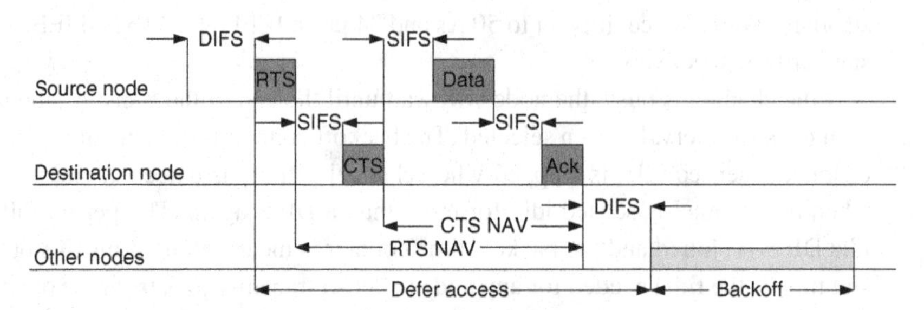

Figure 6.3 RTS/CTS mechanism in WLAN networks using DCF.

which is equal to 2304 octets. Similar to DCF, PCF uses ARQ to increase network reliability.

Furthermore, the request-to-send (RTS)/clear-to-send (CTS) mechanism may be optionally applied to solve the hidden node problem. Note that the RTS/CTS mechanism might be used for frame sizes exceeding a predefined threshold value. Specifically, when a source node is ready for transmission, it performs the aforementioned backoff technique and then transmits an RTS message to the destination node, while the destination node responds to that with a CTS message after waiting for an SIFS. The RTS and CTS frames carry the information of the length of the frame to be transmitted, which is used to update the network allocation vector (NAV) at other nodes. The NAV contains the information about the period of time during which the channel will remain busy. Hence, a node can defer its transmission by detecting either the RTS or CTS frame in order to avoid collisions. When the sender receives the CTS frame, it will start the transmission immediately for a period of time declared in the CTS frame. Figure 6.3 shows the operation of the RTS/CTS mechanism in WLAN networks using DCF.

6.2 QoS in WLAN

While both DCF and PCF offer only best-effort service, the amendment IEEE 802.11e is designed to provide quality-of-service (QoS) support for end-users. In IEEE 802.11e, the hybrid coordination function (HCF) is used for providing parameterized (i.e., absolute) and prioritized (i.e., relative) QoS. HCF uses a contention-based channel access method, referred to as enhanced distributed channel access (EDCA), that operates concurrently with the polling-based HCF-controlled channel access (HCCA) method (Ni [2005]). In both EDCA and HCCA, the so-called transmission opportunity (TXOP) is the main new MAC enhancement of HCF. In EDCA, an STA and, in HCCA, an AP generate the TXOP to allow a given STA to transmit a burst of data frames, separated only by an SIFS from each other. More specifically, IEEE 802.11e divides the channel access time into periodic intervals, called beacon intervals, where a beacon interval consists of the two following periods: (*i*) contention-free period (CFP) and (*ii*) contention period (CP) (Ni [2005]). CFP is used in HCCA for polling the STAs by the AP, while CP is considered in EDCA.

Table 6.2. Traffic class mapping in EDCA.

User priority	IEEE 802.1D designation	AC	IEEE 802.11e designation
1	Background	$AC[BK]$	Background
2	—	$AC[BK]$	Background
0	Best Effort	$AC[BE]$	Best Effort
3	Excellent Effort	$AC[BE]$	Best Effort
4	Controlled Load	$AC[VI]$	Video
5	Video	$AC[VI]$	Video
6	Voice	$AC[VO]$	Voice
7	Network Control	$AC[VO]$	Voice

The IEEE standard 802.11e does not specify any particular scheduling algorithm and allows scheduling to be done in two ways: inter-STA scheduling and intra-STA scheduling. Inter-STA scheduling arbitrates the transmissions of different STAs, while intra-STA scheduling arbitrates the transmissions of different traffic classes in each STA. There are two possible implementation methods. Either inter-STA scheduling is implemented at the AP or TXOP holder, and each STA performs its own intra-STA scheduling, or both inter-STA and intra-STA scheduling are implemented at the AP or TXOP holder. During a TXOP, a single block acknowledgment request (BAR) frame is transmitted at the end of each data transmission by the TXOP holding STA. The receiving STA replies with a block acknowledgment (BA) frame to confirm the frames that were correctly received. Moreover, IEEE 802.11e is able to provide direct connections between STAs by using the so-called direct link protocol (DLP). In a direct connection, the AP can monitor and control the involved STAs by sending beacons periodically. In the following, we describe EDCA and HCCA in greater detail.

6.2.1 EDCA

In IEEE 802.11e QoS-enabled WLANs, EDCA is designed to enhance the contention-based DCF. EDCA deploys four different access categories (ACs), each associated with a different channel access priority. The ACs are deployed using first-in first-out (FIFO) queues, where each data packet is buffered based on its priority. Table 6.2 shows how the eight traffic classes of IEEE 802.1D are mapped to the ACs of EDCA.

To provide relative QoS, each AC queue operates as a single DCF queue, while its channel access parameters are set based on its buffered traffic priority. For instance, for a given queue AC, the following parameters are defined and are broadcast by the AP periodically in each beacon frame: $CW_{min}[AC]$, $CW_{max}[AC]$, $AIFS[AC]$, and $TXOP_{limit}[AC]$. In EDCA, DIFS is replaced by the arbitrary interframe space (AIFS), which generates a random interframe space for access to the channel. The AIFS of a given queue AC is set by $AIFS[AC] = SIFS + AIFSN[AC] \times t_{slot}$, where $AIFSN[AC]$ is the AIFS number and is defined based on the priority of the

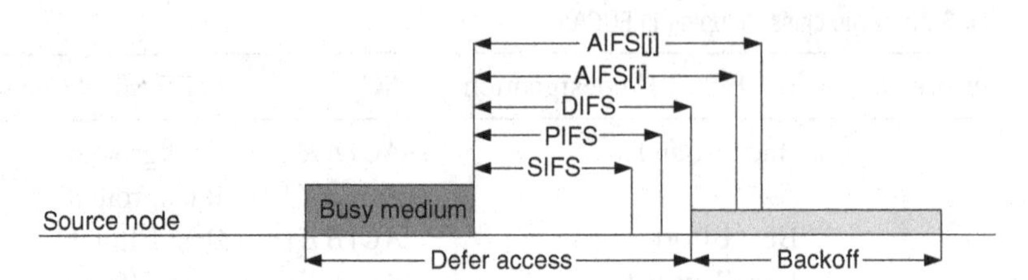

Figure 6.4 Interframe space relationship in QoS-enabled WLAN. After IEEE P802.11e (2005). ©2005 IEEE.

data packet considered for transmission. Note that t_{slot} denotes the time slot of the WLAN network, which is set to 20 μs and 9 μs in IEEE 802.11b and IEEE 802.11a/g, respectively. Similar to DCF, the source node senses the idle medium for a time interval of $AIFS[AC]$ and calculates a random backoff time for each AC using $CW_{min}[AC]$ and $CW_{max}[AC]$. Note that a $TXOP_{limit}[AC]$ is used to bound the transmission time of each AC (Ni [2005]). Figure 6.4 shows the relationship between the different types of interframe space for two access categories $AC[i]$ and $AC[j]$ in a QoS-enabled WLAN. In this figure, the source node senses the idle channel for its corresponding interframe space interval. Then it selects the time slot and decrements its backoff counter as long as the medium is idle. In Fig. 6.4, we consider $AIFS[i] < AIFS[j]$, which indicates that $AC[i]$ has higher priority to access the channel than $AC[j]$.

6.2.2 HCCA

HCCA is an extension of the contention-free PCF technique using the eight traffic classes of IEEE 802.1D in order to provide absolute QoS. In HCCA, STAs report their QoS requirements for each traffic class to the AP by means of traffic specification (TSPEC) frames. A TSPEC frame describes the traffic characteristics and QoS requirements for a given traffic flow, also referred to as traffic stream (TS), including minimum and maximum service interval, minimum data rate, and delay bound. A traffic identifier (TID) is applied by the MAC layer of the end-users to enable the MAC entity to distinguish MSDUs and support QoS differentiation via different service classes. There are sixteen possible TID values, whereby eight identify traffic classes and eight identify parameterized TSs.

In HCCA, the AP is allowed to transmit several contention-free bursts after sensing the channel idle for a PIFS time interval. Note that in IEEE 802.11e, PIFS is defined to be shorter than DIFS and AIFS. As a result, the AP has a higher priority than STAs to access the channel.

In the IEEE standard 802.11e, a simple scheduling algorithm is recommended for AP, where a TS is first established before any data transmission (IEEE P802.11e [2005]). In this algorithm, each STA is allowed to request scheduling of up to eight TSs with different priorities. To initiate a TS connection, the AP gathers the QoS bandwidth request frames (i.e., TSPEC frames) and polls the STAs based on various criteria, e.g., delay requirements of TSs. The AP schedules the STAs by computing the TXOP values according to the QoS requests specified in their TSPEC frames.

6.3 HT WLAN

Applying OFDM and multiple input multiple output (MIMO) antennas in the PHY layer of next-generation IEEE 802.11n WLANs provides various capabilities, such as antenna diversity (selection) and spatial multiplexing. In addition, using multiple antennas provides multipath capability, which increases both throughput and transmission range. The enhanced PHY layer applies two adaptive coding schemes: space time block coding (STBC) and low density parity check coding (LDPC). IEEE 802.11n is able to co-exist with IEEE 802.11 legacy standards, but in greenfield deployments of next-generation high-throughput WLANs (HT WLANs) it is possible to increase the channel bandwidth from 20 MHz to 40 MHz via channel bonding, resulting in significantly increased raw data rates of up to 600 Mb/s. Note that an IEEE 802.11n HT WLAN provides QoS support for end-users by using the above discussed techniques specified in IEEE 802.11e.

6.3.1 Frame aggregation

Due to the PHY and MAC enhancements of IEEE 802.11n, next-generation HT WLANs offer a throughput of at least 100 Mb/s measured at the MAC service access point (SAP). The major MAC enhancement of 802.11n is frame aggregation. As shown in Fig. 6.5, the frame aggregation in HT WLANs comes in two flavors (Skordoulis *et al.* [2008]):

- **A-MSDU:** Aggregate MAC service data unit (A-MSDU) is used to join multiple MSDU subframes (see Fig. 6.5(a)). Specifically, an STA packs multiple MSDUs with possibly different destination addresses into one MAC protocol data unit (MPDU) and sends it to the AP. An AP is allowed to pack multiple MSDUs into one MPDU, whereby all constituent MSDU subframes must be destined to the same STA. In either direction, all constituent MSDU subframes must have the same TID value (i.e., same QoS level).
- **A-MPDU:** Aggregate MAC protocol data unit (A-MPDU) is used to join multiple MPDU subframes (see Fig. 6.5(b)). Specifically, multiple MPDUs with the same receiver address are packed into one PHY service data unit (PSDU). Aggregation of multiple subframes with different TID values in one MPDU is allowed by using multi-TID block acknowledgment (MTBA).

It is important to note that both A-MSDU and A-MPDU require only a single PHY preamble and PHY header. In A-MSDU, the PSDU includes a single MAC header and frame check sequence (FCS), as opposed to A-MPDU where each MPDU subframe contains its own MAC header and FCS. Due to the resultant lower overhead, A-MSDU is able to achieve a higher throughput than A-MPDU for error-free channels. For error-prone channels, however, the throughput of A-MSDU decreases quickly, which

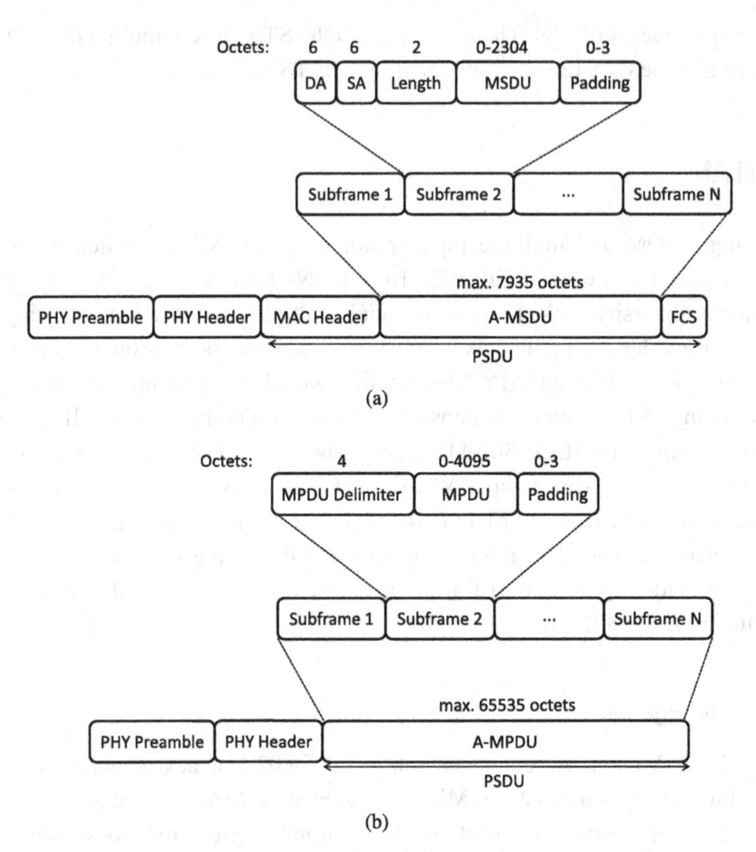

Figure 6.5 Frame aggregation schemes in next-generation WLAN: (a) A-MSDU, and (b) A-MPDU.

indicates the lower robustness of A-MSDU. Thus, adaptively using A-MSDU under good channel conditions and A-MPDU under bad channel conditions yields better performance (Lin and Wong [2006]).

Note that A-MPDU and A-MSDU can be used separately or jointly. It was shown in (Skordoulis *et al.* [2008]) that joint two-level aggregation is able to achieve a higher throughput efficiency. In this approach, the maximum size of A-MSDU is restricted to 4095 octets. Finally, it is worthwhile to mention that fragmentation of MSDUs and A-MSDUs with the same receiver address is allowed.

6.3.2 Reverse direction protocol

The IEEE standard 802.11n introduces the reverse direction (RD) protocol, which provides bidirectional TXOP connections for HT WLANs. Note that the RD protocol might be deployed between an AP and an STA or between a pair of STAs (similar to the DLP in IEEE 802.11e). More precisely, during an RD exchange sequence, the RD initiator can transmit PHY protocol data units (PPDUs) and obtain response PPDUs from the RD responder in a single TXOP. In the HT control field of an IEEE 802.11n PPDU, the

RD grant (RDG)/More PPDU field is used to indicate the RD permission and the last transferred PPDU. In EDCA, an RD responder must transmit the same type of AC data frames as received, while in HCCA an RD responder is allowed to transmit data frames of any TID.

6.3.3 Bandwidth efficiency techniques

In IEEE 802.11n, if a TXOP holding STA that gains access to the channel using EDCA, runs out of frames it can transmit a contention-free end (CF-End) frame to truncate the TXOP and thereby improve bandwidth efficiency. In HT greenfield deployments, the bandwidth efficiency can be further improved by means of a reduced interframe space (RIFS) (2 μs), which cuts down the dead time between frame transmissions. RIFS may be used instead of SIFS to separate multiple transmissions from a single source node. Note that RIFS cannot be applied for SIFS-separated response transmissions, e.g., between data and Ack frames.

In HCCA, the power save multi-poll (PSMP) is one of the new MAC layer features of next-generation WLANs to improve their bandwidth efficiency. The AP sends a PSMP frame in order to schedule the upstream and downstream transmissions of STAs, referred to as uplink transmission time (UTT) and downlink transmission time (DTT), respectively. In a PSMP frame, the PSMP-UTT and PSMP-DTT fields indicate the duration of the uplink and downlink streams dedicated to each STA. The PPDUs transmitted between PSMP-UTT and PSMP-DTT are separated by an SIFS, whereby the PPDUs transmitted within them are separated by an RIFS or an SIFS. The AP monitors all data transmissions in the PSMP-UTT period and can transmit a PSMP recovery frame during a PSMP-UTT, if it detects that the medium is idle and there is enough time in the dedicated PSMP-UTT to transmit the PSMP recovery frame. A PSMP recovery frame consists of a modified PSMP-UTT (and/or PSMP-DTT) for the currently scheduled STA and PSMP-UTTs for other STAs that were originally scheduled after this PSMP-UTT. Note that the AP should not modify the schedules of other STAs (IEEE P802.11n [2009]).

Moreover, HT WLANs support a PSMP burst transmission mode. After transmitting an initial PSMP sequence, additional PSMP sequences can be transmitted by the AP in order to allow for a more efficient resource allocation and error recovery (IEEE P802.11n [2009]). The PSMP burst consists of data and MTBA frames, where

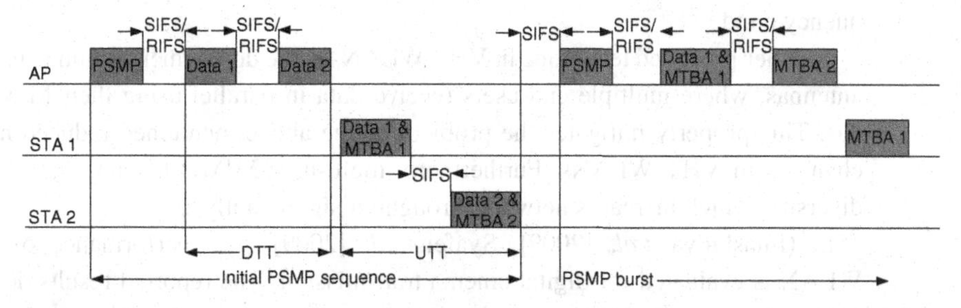

Figure 6.6 PSMP burst transmission in a HT WLAN. After IEEE P802.11n (2009). ©2009 IEEE.

each data frame may include multiple packets with different TIDs aggregated through A-MPDU. Figure 6.6 illustrates the PSMP burst transmission in an HT WLAN. In this figure, each PSMP frame has a 'More PSMP' flag bit to indicate the status of the burst. More specifically, the burst transmission remains active until the "More PSMP" bit of the last PSMP sequence is set to 0 in order to denote the end of the burst.

6.4 VHT WLAN

In 2007, the IEEE 802.11 very high-throughput (VHT) WLAN study group was set up to introduce next-generation Gigabit VHT WLANs (IEEE P802.11 VHT [2010]). In 2008, the VHT study group formed the two following VHT task groups: (*i*) VHTL6: IEEE 802.11ac (IEEE P802.11ac VHT [2010]) to operate at a frequency band below 6 GHz, and (*ii*) VHT60: IEEE 802.11ad (IEEE P802.11ad VHT [2010]) to operate in the 60 GHz millimeter-wave (mm-wave) frequency band.

One of the major challenges of the IEEE 802.11ac task group is providing backward compatibility with legacy IEEE 802.11a/b/g WLANs as well as HT WLANs operating in the 2.4 GHz and 5 GHz frequency bands. On the other hand, the major challenging issue of the IEEE 802.11ad task group is the deployment of the 60 GHz mm-wave frequency band given its coverage limitations and huge energy consumption. Both VHTL6 and VHT60 aim at supporting future bandwidth-hungry applications, e.g., high-definition (HD) video streaming (Jeon [2008], Grodzinsky and de Vegt [2008], Imashioya *et al.* [2009]).

6.4.1 VHTL6

IEEE 802.11ac VHT benefits from the advanced technologies applied in IEEE 802.11n HT WLANs, such as OFDM and MIMO antennas. Advanced modulation schemes play a key role in increasing the bit rate of VHT WLANs. Specifically, 256-quadrature amplitude modulation (QAM) is considered in VHT WLANs instead of the 64-QAM of HT WLANs.

In VHT WLANs, it is possible to increase the channel bandwidth from 40 MHz of HT WLANs to 80 MHz and 160 MHz via channel bonding, resulting in significantly increased raw data rates of up to 1 Gb/s. Note that this bandwidth improvement results in a reduced number of non-overlapping channels in the considered unlicensed frequency band.

Another proposed technique in VHT WLANs is the deployment of multi-user MIMO antennas, where multiple end-users receive data in parallel using their MIMO antennas. This property mitigates the problem of the above-mentioned reduced number of channels in VHT WLANs. Furthermore, multi-user MIMO antennas provide spatial diversity, which increases network throughput significantly.

In (Imashioya *et al.* [2009], Syafei *et al.* [2009]), the performance of VHTL60 WLAN is evaluated for digital cinema transmission. The reported results demonstrate the feasibility of achieving 1.2 Gb/s throughput over a distance of 33 m by utilizing an 80 MHz channel in the 5 GHz frequency band.

6.4.2 VHT60

IEEE 802.11ad VHT makes use of the unlicensed 57–66 GHz frequency band, which has the potential to provide multi-Gb/s wireless communications for VHT WLAN end-users. Toward the deployment of a 60 GHz IEEE 802.11ad VHT WLAN various MAC and PHY layer modifications are needed, including MAC layer modifications for directional antennas, fast session transfer between PHY layers, beamforming, and spatial reuse (Perahia *et al.* [2010]).

In a VHT60 WLAN, AP and STAs are equipped with 60 GHz and 2.4/5 GHz devices. The IEEE 802.11ad task group addresses the two following operational requirements of VHT60 WLAN networks (Perahia *et al.* [2010]): (*i*) backward compatibility with legacy IEEE 802.11 WLAN infrastructure networks in terms of management plane, e.g., association, authentication, and security; and (*ii*) fast session transfer between PHY layers, which provides seamless rate fallback among VHT and IEEE 802.11n multi-band end-users.

The MAC layer of a VHT60 network is based on EDCA and HCCA, whereby PSMP is considered to improve the energy consumption of STAs (Nandagopalan *et al.* [2009]). Note that the energy consumption is a critical issue in VHT60 WLAN networks due to their transmissions in the 60 GHz frequency band.

6.4.3 VHT applications

Wireless sensor networks are becoming ubiquitous in our daily activities, e.g., security surveillance and environmental monitoring. Wireless sensor networks allow for their flexible and cost-efficient installation and maintenance, giving rise to a number of interesting applications (Li [2009]). For instance, one of the most popular aeronautic research topics is the avionic structural health monitoring system using HD video sensor networks (Kuehl [1996], Lamberth *et al.* [2003], Jian *et al.* [2010]).

VHT60 with its huge bandwidth is a promising solution for wireless sensor networks that require very high throughput operation for sensing, processing, transmitting, and actuating functions (Jian *et al.* [2010]). Because of its flexible frequency reuse, superior transmission capacity, and huge amount of available bandwidth, VHT60-based wireless sensor networks can provide bit rates of up to 40 Gb/s for very high throughput sensing and actuating applications (Chang *et al.* [2010]).

7 WiMAX

Worldwide interoperability for microwave access (WiMAX) has been envisioned by the WiMAX forum as a single worldwide adopted standard for high-speed wireless metropolitan area networking. The term WiMAX denotes wireless metropolitan area network (WMAN) technology based on IEEE 802.16 specifications. In this chapter, we provide an overview of the salient features and most important specifications of legacy and next-generation WMANs.

7.1 Fixed WiMAX

The initial IEEE 802.16 WiMAX standard was established in the frequency band 10–66 GHz, providing up to 75 Mb/s line-of-sight (LOS) connections for both point-to-multipoint and mesh modes. Table 7.1 summarizes the IEEE 802.16 WiMAX standard family.

7.1.1 PHY layer

IEEE 802.16a provides non-LOS connections in the frequency band of 2–11 GHz (licensed and unlicensed). The WiMAX physical (PHY) layer supports the following four different modulation schemes: WirelessMAN-SC (single carrier), WirelessMAN-SCa (single carrier access), WirelessMAN-OFDM (orthogonal frequency division multiplexing), and WirelessMAN-OFDMA (orthogonal frequency division multiple access). While WirelessMAN-SC has been designed for the frequency band 10–66 GHz, the other modulation schemes can be used for the frequency band 2–11 GHz. Additionally, the WiMAX PHY layer transfers bidirectional data by means of time division duplex (TDD) or frequency division duplex (FDD).

7.1.2 MAC layer

IEEE 802.16 is a connection-oriented standard, i.e., prior to transmitting data between subscriber stations (SSs) and base station (BS) connections must be established. Each connection is identified by a 16-bit connection identifier (CID). The WiMAX medium access control (MAC) layer is responsible for assigning CIDs as well as allocating bandwidth to SSs. It consists of the following three sublayers: (*i*) convergence sublayer (CS),

Table 7.1. IEEE 802.16 WiMAX standard family.

Standard/Amendment	Description
802.16	Up to 75 Mb/s, 10–66 GHz
802.16a	Up to 75 Mb/s, 2–11 GHz
802.16d	Fixed WiMAX, 2–11 GHz
802.16e	Mobile WiMAX, up to 75 Mb/s, lower than 6 GHz
802.16f	Management information base (MIB) enhancements
802.16g	Management plane procedures and services
802.16h	Coexistence mechanisms for unlicensed frequency bands
802.16i	Mobile management information base
802.16j	Multihop relay WiMAX
802.16k	WiMAX bridging
802.16m	Gigabit WiMAX
802.16n	Reliability enhancements

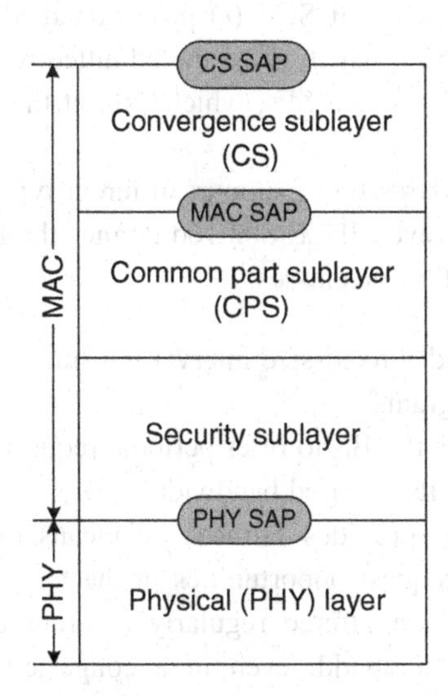

Figure 7.1 IEEE 802.16 WiMAX reference model. After Li *et al.* (2007). ©2007 IEEE.

whereby different higher-layer protocols are implemented in different CSs, e.g., ATM CS and packet CS are used for ATM and Ethernet networks, respectively; (*ii*) common part sublayer (CPS), which is responsible for bandwidth allocation and generating MAC protocol data units (MPDUs); and (*iii*) security sublayer (Kuran and Tugcu [2007]). Figure 7.1 shows the IEEE 802.16 WiMAX reference model with the aforementioned PHY layer and three MAC sublayers.

To access the channel, each SS must send an uplink bandwidth request (UL-request) to the BS and the BS responds with an uplink bandwidth grant (UL-grant) subsequently. The UL-grant services assign bandwidth based on the quality-of-service (QoS) parameters of that connection. Although bandwidth requests are per connection, the BS does not allow the entire uplink capacity to be granted to a single SS. When a UL-grant is sent by the BS, the receiver SS cannot change or share the size of the granted bandwidth (Cicconetti *et al.* [2006]). In IEEE 802.16, there are two modes of transmitting UL-requests: (*i*) contention mode and (*ii*) contention-free mode (polling). In the contention mode, SSs send their UL-requests during the contention period, whereby contention is resolved by means of back off. In the contention-free mode, the BS polls each SS and SSs reply by sending UL-requests. The downlink subframe structure of IEEE 802.16 WiMAX is shown in Fig. 7.2. In this figure, the downlink burst profiles are tagged with downlink interval usage codes (DIUCs). Similarly, the uplink subframes consist of uplink interval usage codes (UIUCs) to indicate uplink transmissions (Eklund *et al.* [2002]).

The requested services of each SS are first registered during the initialization phase and connections are established subsequently. If a given SS changes its services, additional connections can be established in the network. Each connection is associated with a service flow (SF). An SF is defined based on available scheduling services and includes a set of QoS parameters, an SF identifier (SFID), and a CID (Li *et al.* [2007]). Generally, WiMAX applies three different types of SFs: (*i*) provisioned SFs, which are represented by an SFID; (*ii*) admitted SFs, which show the admitted requests for the available resources and/or SFs; and (*iii*) active SFs, which exhibit the allocated resources and/or SFs.

No specific scheduling algorithm is standardized to support different types of data flows and QoS for fixed WiMAX SSs. However, IEEE 802.16d defines the following four scheduling services to support different traffic classes:

1. **Unsolicited grant service (UGS):** provides fixed-size interval transmission opportunities without the need for requests or grants.
2. **Real-time polling service (rtPS):** allows the BS to offer periodic request opportunities to a specific SS in order to indicate its required bandwidth.
3. **Non-real-time polling service (nrtPS):** provides fairness by means of allocating the contention and regular unicast request opportunities in the UL-request of each SS. Unicast request opportunities are offered regularly in order to ensure that the SS has a chance to request bandwidth even in a congested network environment.
4. **Best effort (BE):** is based on contention and provides non-regular unicast request opportunities.

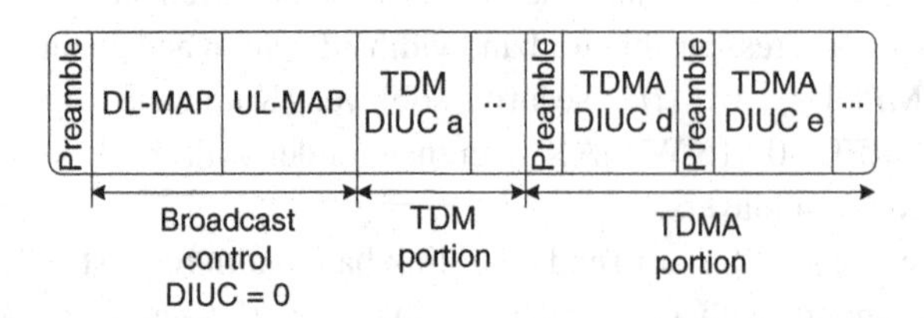

Figure 7.2 Downlink subframe structure of IEEE 802.16 WiMAX. After Eklund *et al.* (2002). ©2002 IEEE.

7.2 Mobile WiMAX

The IEEE standard 802.16e, also known as mobile WiMAX (Vaughan-Nichols [2008]), offers scalability with regard to both radio access technology and network architecture. While the spectrum allocation is applied as a radio access technology in mobile WiMAX, its flexibility in network deployment provides various services. For instance, it is able to provide wireless Internet and support seamless IP mobility for end-users (Sim *et al.* [2009]). Based on this property of mobile WiMAX, the International Telecommunication Union (ITU) adopted mobile WiMAX as one of the International Mobile Telecommunications (IMT) technologies (Wang *et al.* [2008]). Multiple input multiple output (MIMO) antenna techniques and flexible sub-channelization schemes are deployed to provide high data rate connections (Li *et al.* [2007]).

Figure 7.3 shows the high-level MAC/PHY protocol structure specified in IEEE 802.16e mobile WiMAX (Etemad [2008]). Similar to the fixed WiMAX MAC layer, the mobile WiMAX MAC layer consists of CS, CPS, and security sublayers. The modules of the CPS sublayer are categorized into mobility control and resource management as well as control and support for the wireless channels in the PHY layer. In Fig. 7.3, the PHY control module handles PHY layer signaling mechanisms, i.e., ranging and link adaptation (e.g., power control). The control signaling block is responsible for generation and transmission of control messages such as resource allocation messages. From another perspective, we may classify the modules of the MAC/PHY protocol into control and data planes based on their functional domains (see Fig. 7.3). As shown in the figure, the automatic response request (ARQ) mechanism can be used in the data plane to enhance network reliability.

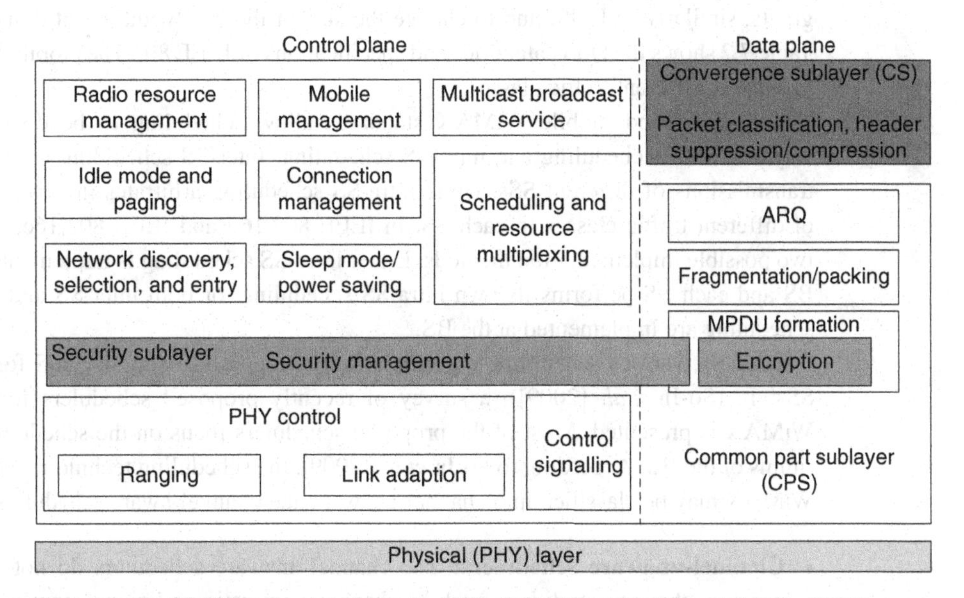

Figure 7.3 MAC/PHY protocol structure in mobile WiMAX. After Etemad (2008). ©2008 IEEE.

Table 7.2. QoS categories and specifications of IEEE 802.16e mobile WiMAX (Li *et al.* [2007]).

QoS category	Applications	QoS specifications
Unsolicited grant service (UGS)	Voice over Internet protocol (VoIP)	Maximum substained rate Maximum latency tolerance Jitter tolerance
Extended real-time polling service (ErtPS)	VoIP with activity detection	Minimum reserved rate Maximum substained rate Maximum latency tolerance Jitter tolerance
Real-time polling service (rtPS)	Audio/Video streaming	Minimum reserved rate Maximum substained rate Maximum latency tolerance
Non-real-time polling service (nrtPS)	File transfer protocol (FTP)	Minimum reserved rate Maximum substained rate
Best effort service (BE)	Web browsing	Maximum substained rate

7.2.1 QoS in mobile WiMAX

Similar to IEEE 802.16d fixed WiMAX, no specific scheduling algorithm is standardized to support QoS for mobile WiMAX SSs. In addition to the four aforementioned scheduling services of fixed WiMAX, IEEE 802.16e mobile WiMAX defines extended-real-time polling service (ErtPS), which is able to offer unsolicited unicast grants, similarly to UGS, and to change the size of the bandwidth grant dynamically. Table 7.2 shows the QoS categories and specifications of IEEE 802.16e mobile WiMAX to support different traffic classes.

Both fixed and mobile WiMAX standards allow scheduling to be done in two ways: inter-SS scheduling and intra-SS scheduling. Inter-SS scheduling arbitrates the transmissions of different SSs, while intra-SS scheduling arbitrates the transmissions of different traffic classes in each SS. In IEEE 802.16d and IEEE 802.16e, there are two possible implementation methods. Either inter-SS scheduling is implemented at the BS and each SS performs its own intra-SS scheduling, or both inter-SS and intra-SS scheduling are implemented at the BS.

Recently, various scheduling algorithms were proposed to support QoS for mobile SSs. In (So-In *et al.* [2009]), a survey of recently proposed schedulers for mobile WiMAX is presented. Most of the proposed schedulers focus on the scheduling algorithms of the BS. According to So-In *et al.* [2009], the scheduling techniques of mobile WiMAX may be classified into channel-unaware and channel-aware schedulers:

- **Channel-unaware schedulers:** The channel-unaware schedulers do not consider wireless channel conditions such as channel error ratio and signal power status in their channel and bandwidth assignments. This type of scheduler is designed to meet the QoS requirements of the five different scheduling services, e.g., in terms of throughput and delay. Typically, the channel-unaware scheduling can be done intra-class and inter-class. Intra-class scheduling is used to allocate the resources to the same traffic class. The round-robin (RR) scheduling algorithm family, including weighted round-robin (WRR), is an example of intra-class scheduling. Inter-class scheduling may be applied for various traffic types with different QoS requirements. This type of scheduler can be viewed as a second-level hierarchical scheduling algorithm, where the intra-class scheduling queues are scheduled based on their QoS requirements. Priority queuing (PQ) is an example of inter-class schedulers where the slot allocation of PQ is suitable for higher-priority queues.

- **Channel-aware schedulers:** Unlike channel-unaware schedulers that assume error-free channel conditions, channel-aware schedulers take the wireless channel conditions (e.g., attenuation and fading) into account in their scheduling and bandwidth assignments. The mobility of SSs in IEEE 802.16e is a challenging issue that has to be considered in channel-aware scheduling algorithms. For instance, it is possible that an SS receives the UL-grants of the BS, but it is not able to transmit during its scheduled uplink period due to high error rates. Channel-aware scheduling algorithms may be classified into the four following classes: fairness, QoS guarantee, system throughput maximization, and power optimization. The fairness channel-aware scheduling algorithm is mostly applied for BE traffic, where fair resource allocation is more important than network throughput. To minimize the end-to-end delay and respect the predefined threshold delay value, the QoS guarantee scheduling algorithm is used, especially for rtPS and nrtPS traffic. In the system throughput maximization scheduling algorithm, the scheduler maximizes the total network throughput by allocating a minimum number of resources (i.e., time slots). The power optimization channel-aware scheduling algorithm attempts to optimize the throughput and power consumption of the network.

7.2.2 Mobile WiMAX handover

One of the important challenges of a mobile WiMAX network is the fast and seamless handovers of mobile SSs. Although IEEE 802.16e standard defines the MAC layer handover management framework, a more detailed study has recently started to develop the upper layers (Ray *et al.* [2010]). The proposed MAC layer handover management framework provides various flexibilities in initiating and optimizing handovers. It supports different handover activities such as intra- and inter-cell handovers.

In the IEEE 802.16e mobility structure, the three following link layer handover mechanisms are proposed: (*i*) hard handover (HHO), (*ii*) macro-diversity handover (MDHO), and (*iii*) fast base station switching (FBSS). In the mobile WiMAX standard, HHO is considered as the default handover mechanism. HHO is a break-before-make (BBM) technique, where the SS should be disconnected before connecting to the next BS with a possible service interrupt. MDHO and FBSS apply the make-before-break (MBB) approach, also known as soft handover, where a given SS communicates with multiple BSs simultaneously. Similar to most of the WMAN technologies, in mobile

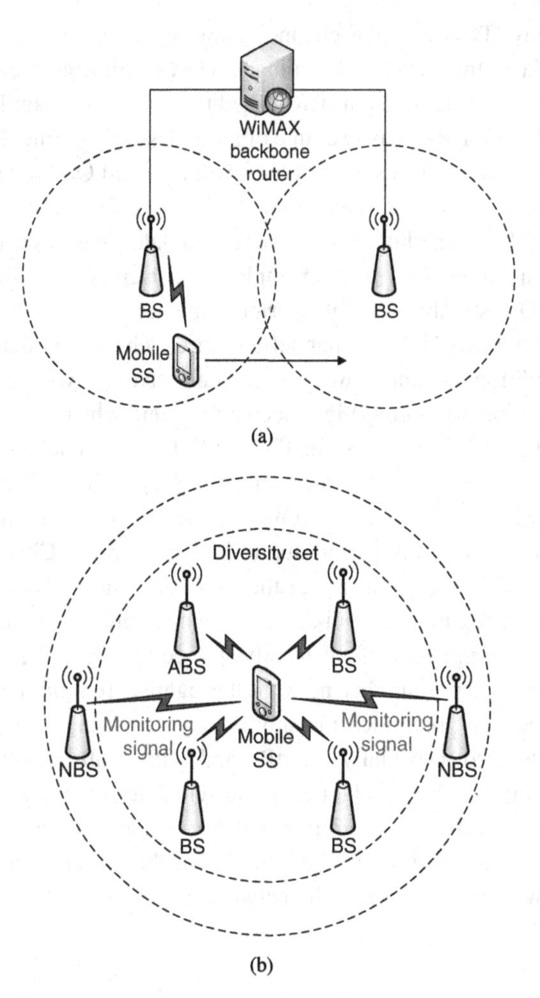

Figure 7.4 Mobile WiMAX handover mechanisms: (a) hard handover (HHO) and (b) macro-diversity handover (MDHO). After Ray *et al.* (2010). ©2010 IEEE.

WiMAX a handover is initiated based on the received signal strengths at the BS or mobile SS. Note that both BS and SS may schedule the handover initiation jointly (Ray *et al.* [2010]).

Figure 7.4 shows the HHO and MDHO procedures proposed in the IEEE 802.16e mobile WiMAX standard (Ray *et al.* [2010]). In the network topology acquisition phase of HHO, the serving BS prepares a list of neighboring BSs (NBSs) using the WiMAX backbone router. In this phase, the serving BS periodically broadcasts the status information of its in-range NBSs. The SS scans the received list of NBSs and transmits a scanning interval allocation request (SCN-REQ) to its serving BS. The BS negotiates with the potential target BS and transmits a scanning interval allocation response (SCN-RSP) to the SS. For synchronization and ranging between the PHY layers of the SS and target BS, the SS transmits a ranging request (RNG-REQ) message to the target BS, which in turn responds by sending a ranging response (RNG-RSP) message.

In the handover phase of HHO, the SS transmits a mobile handover indication (MOB HO-IND) message to inform the serving BS about the handover status. Subsequently, the SS terminates its connection with the serving BS and starts the synchronization and ranging process to resume downlink/uplink retransmissions with the target BS. Note that the actual handover phase of HHO consists of the authorization and registration phases, which can be a time-consuming process.

In the soft-handover techniques (i.e., MDHO and FBSS), a diversity set (DS) and anchor BS (ABS) are defined to perform the fast and seamless handover. As shown in Fig. 7.4(b), each SS has a DS area, which indicates the set of BSs in the range of SS that is involved in the handover process. The BSs are selected based on their signal strengths, while the BS with the most powerful signal strength is called an ABS. In MDHO, the SS simultaneously communicates with all the BSs of its DS. As shown in the figure, the SS continuously monitors the signal strengths of its NBSs for efficient updating of its DS and ABS. The handover occurs once the status of the BSs in the DS of the SS is changed or an NBS is added to the DS area. In the FBSS mechanism, the SS communicates only with the ABS. However, it continuously monitors the signal strengths of the NBSs and the other BSs of its DS (Ray *et al.* [2010]).

The mobile WiMAX HHO is simpler and more cost-efficient than the soft-handover schemes. However, it is not a good approach for voice applications (e.g., VoIP) due to its communication interruptions, which result in an increased packet loss. Instead, by using MDHO and FBSS schemes the mobile SS experiences a very low packet loss (less than 1%) and handover latency (less than 50 ms) for high-speed activities (up to 120 km/h) (Ray *et al.* [2010]). These handover techniques support high-speed real-time VoIP applications by deploying a large number of BSs at the expense of significantly increased deployment costs.

7.3 Next-generation WiMAX

7.3.1 Multihop relay WiMAX

To extend the coverage of WMAN networks, one option might be to increase the cell size by using powerful antennas. This approach cannot provide fair service to end-users since the SSs closer to the BS receive a stronger signal and better service than SSs further away from the BS. Another approach might be to increase the number of wired BSs, which is costly since for each BS a wired cable has to be deployed. Conversely, multihop relaying is a promising approach for WMANs to extend network coverage in a cost-efficient manner.

Some studies have been carried out to show the feasibility of deploying multihop relay WiMAX networks, e.g., (Lu *et al.* [2008]). In 2009, the IEEE 802.16j amendment was approved by the IEEE standards board in order to add multihop relay capabilities to single-hop IEEE 802.16e mobile WiMAX.

An important challenge of IEEE 802.16j is compatibility with point-to-multipoint OFDMA-based IEEE 802.16e networks and SS equipment (Peters and Heath [2009]).

In a multihop relay WiMAX network architecture, SSs communicate with the BS through the relay stations (RSs). To deploy multihop relay WiMAX, IEEE 802.16j proposes various modifications in the PHY and MAC layers of IEEE 802.16e BSs (Peters and Heath [2009]). To accommodate the relay capability, each uplink/downlink subframe is split into an *access zone* and a *relay zone*. While an RS uses the access zone to communicate with the SSs, the relay zone is used to communicate with another RS or BS. In addition to legacy single-channel transmissions of BS/RSs, multi-channel transmission is proposed to increase network throughput. Moreover, the IEEE standard 802.16j provides path diversity by means of cooperative relaying between RSs. The following three mechanisms are proposed to implement cooperative path diversity (Peters and Heath [2009]): (i) cooperative source diversity, where an RS transmits identical signals in both time and frequency domains, (ii) cooperative transmit diversity, where transmissions are done using space-time codes, and (iii) cooperative hybrid diversity, which applies space-time codes to transmit across a subset of RSs and cooperative source diversity is used among SSs.

In the MAC layer design of multihop relay WiMAX networks, some functions are dedicated to the RSs in order to deploy a distributed scheduling algorithm for their local traffic among SSs and a centralized scheduling algorithm for relaying traffic. Recently, various centralized and distributed scheduling algorithms have been proposed for multihop relay WiMAX networks. For instance, a joint routing–scheduling algorithm for TDMA-based IEEE 802.16j multihop relay WiMAX was proposed in (Hong and Pang [2009]), where both interference-aware routing and link scheduling are considered to maximize network throughput. The reported results demonstrate that applying a joint routing and scheduling algorithm outperforms disjoint routing and/or scheduling algorithms in terms of network throughput.

In an IEEE 802.16j-based multihop relay WiMAX network, each RS defines and uses two CIDs: a local CID and a relay CID. It is important to note that the RS is responsible for assigning the relay CIDs and must avoid relay CID duplications. IEEE 802.16j introduces a new MAC protocol, called relay-MAC (R-MAC), to deploy connection tunneling between RSs and BS (Peters and Heath [2009]). In R-MAC, each tunnel connection is identified by a tunnel CID (T-CID) between an RS and BS to carry the MPDUs of multiple SSs' connections. Moreover, the control and management packets are carried through the management tunnel connections. Note that in each T-CID the connections must have the same traffic class.

In addition to scheduling and resource management, the placement of BS and RSs plays a key role in the performance of the network. A relay-centric hierarchical optimization model for joint radio resource management and network planning of a multihop relay WiMAX network was proposed in (Niyato *et al.* [2009]). Based on the location of the BS and RSs, the model optimizes the amount of reserved bandwidth on each wireless relay link and performs admission control for each SS with the objective of maximizing the utility of RSs. The numerical results show that the proposed joint model outperforms static radio resource management schemes.

In the case where the location of the BS and RSs are fixed and known to mobile SSs, an optimal RS selection scheme for vehicular SSs was presented in (Ge *et al.* [2010]).

The proposed selection scheme is designed to maximize end-to-end network capacity by using nonlinear optimization techniques. The numerical results show that the proposed approach is able to increase the end-to-end network capacity significantly.

One of the major challenging issues in IEEE 802.16j-based multihop relay WiMAX networks are the frequent handovers. In the IEEE standard 802.16j, RS grouping is proposed as an optional mechanism to decrease the number of handovers. In the RS grouping technique, an RS group consists of a number of neighboring RSs that form a logical RS with a larger coverage. In (Yang *et al.* [2010]), a new RS grouping algorithm was designed to minimize handovers by applying a greedy grouping policy, where RSs with higher handover rates are selected to form an RS group. The reported simulation results show that the number of handovers decreases significantly by using the proposed RS grouping algorithm. Moreover, it was shown that a small RS group size increases the throughput of the network with fixed SSs, while a large RS group size should be selected for mobile SSs (Yang *et al.* [2010]).

7.3.2 Gigabit WiMAX

Recently, the IEEE 802.16m task group (IEEE P802.16m/D5 [2010]) was formed to design next-generation Gigabit, also known as fourth-generation (4G), WiMAX networks. A Gigabit WiMAX network is based on IEEE 802.16e mobile WiMAX and emerging IEEE 802.16j multihop relay WiMAX to provide high-speed wireless Internet access for end-users. Gigabit WiMAX supports both fixed and mobile SSs by providing up to 1 Gb/s and 100 Mb/s data rates, respectively. Backward compatibility with the legacy WiMAX technologies is one of the important goals in the design of Gigabit WiMAX. Gigabit WiMAX is compatible with all OFDM-based mobile broadband access technologies. This property provides enormous flexibility and evolutionary upgrade possibilities for both service providers and end-users. By means of interworking functions, Gigabit WiMAX enables roaming and seamless connectivity across IMT-Advanced and IMT-2000 systems (Ahmadi [2009]). Note that the requirements of IMT-Advanced networks (ITU-R M.2134 [2008]) are considered in IEEE 802.16m.

Table 7.3 summarizes the major features of IEEE 802.16m Gigabit WiMAX networks. Various PHY and MAC layer enhancements are considered in the standard (Papapanagiotou *et al.* [2009]). IEEE 802.16m defines a new frame structure in order to offer flexibility in allocating slots for upstream and downstream transmissions. It specifies a 20-ms superframe, which is divided into 5-ms radio frames using either TDD or FDD. By means of OFDMA, it is possible to divide each 5-ms frame into eight subframes, whereby each subframe is used for either downlink or uplink transmissions. The use of space-time coding techniques allows MIMO antennas of Gigabit WiMAX networks to support up to eight stream transmissions at the BS. These techniques help increase data rates and network reliability significantly. IEEE 802.16m supports multiuser MIMO antennas, where multiple SSs receive data in parallel using their MIMO antennas (Li *et al.* [2009, 2010a]). Multicarrier transmission is another key technology

Table 7.3. Features of IEEE 802.16m Gigabit WiMAX (Papapanagiotou *et al.* [2009]).

Feature	IEEE 802.16e	IEEE 802.16m
Aggregate data rate	75 Mb/s	100 Mb/s for mobile SSs 1 Gb/s for fixed SSs
Frequency band	Lower than 6 GHz	Lower than 6 GHz
Duplexing scheme	TDD and FDD	TDD and FDD
MIMO support	Up to 4 streams	Up to 8 streams
Coverage	Up to 10 km	Up to 100 km
Handover inter-frequency interruption time	35–50 ms	30 ms
Handover intra-frequency interruption time	Not specified	100 ms
Mobility speed	Up to 120 km/h	Up to 350 km/h
IDLE to ACTIVE state transition	390 ms	50 ms
QoS	5 scheduling services	5 scheduling services

to increase data rates in Gigabit WiMAX networks, whereby parallel transmissions on multiple radio frequency (RF) carriers are allowed (Fu *et al.* [2010]).

In the MAC layer of Gigabit WiMAX, multihop relay functions are defined to increase network coverage and resource allocation efficiency. Moreover, the MAC layer controls the PHY over multiple frequencies and multi-user MIMO antennas. It supports fast session transfer between PHY layers, beamforming, and spatial reuse. The IEEE 802.16m MAC layer also deploys state-of-the-art power saving and handover techniques to provide mobility support for high-speed mobile SSs (Papapanagiotou *et al.* [2009]).

Two advanced power conservation mechanisms, i.e., sleep and idle modes, are proposed to decrease signaling and operational overhead as well as to provide higher power saving gains than existing power management techniques (Kim and Mohanty [2010]). Typically, a mobile SS operates in one of the following modes based on given traffic conditions: (*i*) connected mode, (*ii*) sleep mode, or (*iii*) idle mode. In next-generation Gigabit WiMAX networks, the sleep mode is activated under light and/or real-time traffic loads in order to perform power saving (especially for mobile SSs). Two power saving classes are introduced for different types of traffic. While non-real-time and BE traffic are considered in one class, another power saving class is defined for real-time traffic (Jin *et al.* [2010]). In sleep mode, a mobile SS operates in the listen and sleep modes. During the listening window, the SS monitors the downlink for a traffic indication message to terminate its sleep mode and move to the connected mode. The sum of listening and sleep windows form a sleep cycle. In IEEE 802.16m Gigabit WiMAX networks, the dynamic adjustment of the sleep cycle is allowed to adapt to current traffic pattern changes. To improve bandwidth efficiency and power saving gains, the listening window can be terminated before the scheduled time via an explicit indication message transmitted by the BS (Kim and Mohanty [2010]). In the idle mode, a mobile SS is disconnected from the BS, but the idle mode retention information (IMRI), e.g., security keys and service flow parameters, are retained at the BS. The IMRI is designed to expedite status changes of the idle mobile SS. Moreover, the idle mobile SS periodically turns on its receiver at pre-specified intervals to monitor paging messages. Consequently, the SS is able to check for pending downlink traffic at the BS (Kim and Mohanty [2010]).

8 LTE

Long term evolution (LTE) has been defined by the third generation partnership project (3GPP) as fourth-generation (4G) cellular network technology for high-speed wireless end-users. In this chapter, we provide an overview of the salient features and most important specifications of LTE and next-generation LTE-Advanced networks.

8.1 PHY layer

The first amendment of LTE (release 8) provides a transmission rate of 300 Mb/s and operates in both time division duplex (TDD) and frequency division duplex (FDD) modes. 4G LTE provides simplicity for both operators and end-users (Pospishny *et al.* [2010]). LTE operators are given the flexibility to define the size of bandwidth, ranging from below 5 MHz up to 20 MHz. Furthermore, various user-friendly features have been considered in 4G LTE networks, including plug-and-play and self-configuration.

LTE aims at providing a smooth evolution from earlier 3GPP and 3GPP2 cellular networks such as wide-band code division multiple access/high-speed packet access (WCDMA/HSPA) and code division multiple access (CDMA2000) (Astely *et al.* [2009]). Typically, orthogonal frequency division multiplexing (OFDM) is used in the downlink radio transmission of LTE networks. Using narrow-band subcarriers in combination with a cyclic prefix leads to a radio transmission that is robust against time dispersion. As a result, the cost and power consumption of mobile end-users decrease due to the simplified receiver baseband processing. Moreover, LTE supports advanced multi-antenna schemes such as single/multiple-user multiple input multiple output (MIMO) antennas, transmit diversity, spatial multiplexing, and beamforming.

From a physical (PHY) layer standardization perspective, the evolutionary upgrade from third generation (3G) to TDD-based International Mobile Telecommunications (IMT)-Advanced systems passes through the 4G LTE technology. More specifically, time division-synchronous code division multiple access (TD-SCDMA) is considered to be used for 3G networks, while its advanced schemes have been proposed for future cellular networks. Figure 8.1 shows the milestones of the TD-SCDMA evolution (Peng *et al.* [2010]). In each phase of Fig. 8.1, different network technologies have been introduced using the released standard draft. For instance, release 5 (R5) specifications consist of some advanced enabling functions, including high-speed downlink packet access (HSDPA), synchronization procedures, and terminal location in the

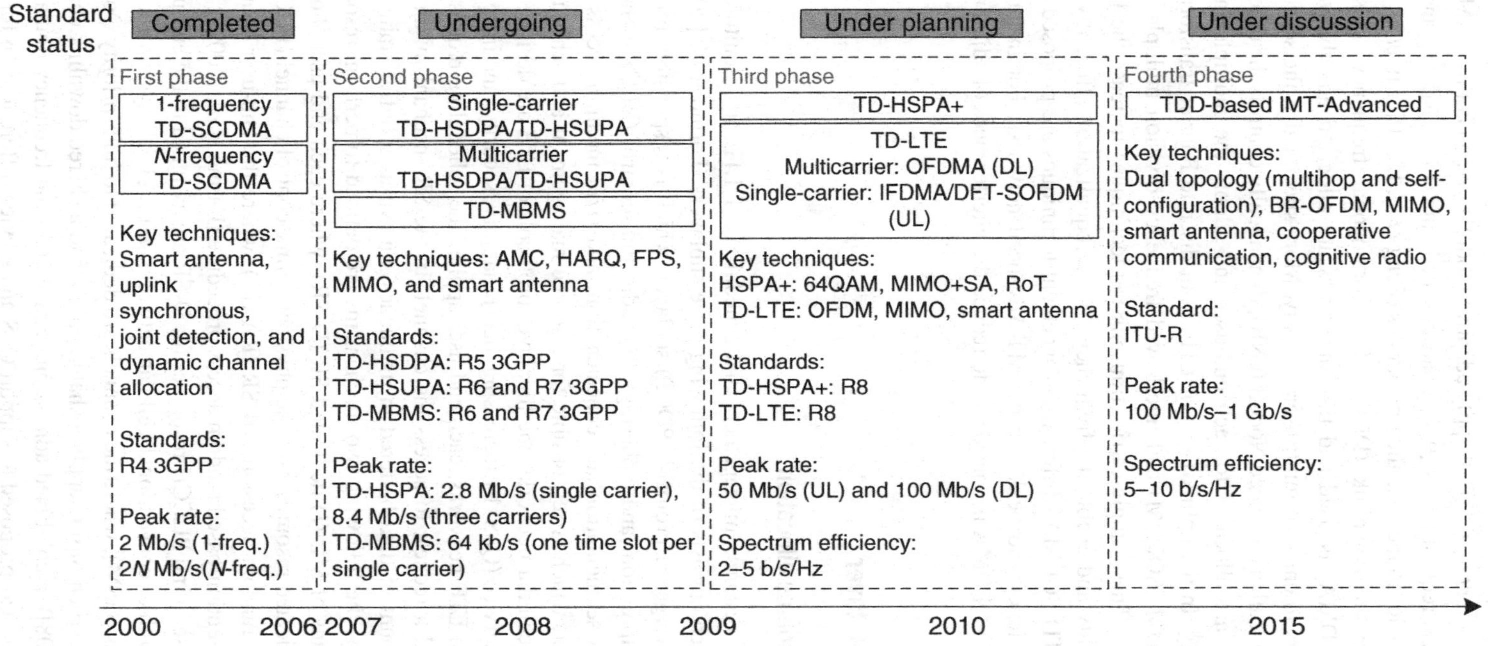

Standard status

Completed

First phase

1-frequency TD-SCDMA

N-frequency TD-SCDMA

Key techniques:
Smart antenna, uplink synchronous, joint detection, and dynamic channel allocation

Standards:
R4 3GPP

Peak rate:
2 Mb/s (1-freq.)
2*N* Mb/s(*N*-freq.)

Undergoing

Second phase

Single-carrier TD-HSDPA/TD-HSUPA

Multicarrier TD-HSDPA/TD-HSUPA

TD-MBMS

Key techniques: AMC, HARQ, FPS, MIMO, and smart antenna

Standards:
TD-HSDPA: R5 3GPP
TD-HSUPA: R6 and R7 3GPP
TD-MBMS: R6 and R7 3GPP

Peak rate:
TD-HSPA: 2.8 Mb/s (single carrier), 8.4 Mb/s (three carriers)
TD-MBMS: 64 kb/s (one time slot per single carrier)

Under planning

Third phase

TD-HSPA+

TD-LTE
Multicarrier: OFDMA (DL)
Single-carrier: IFDMA/DFT-SOFDM (UL)

Key techniques:
HSPA+: 64QAM, MIMO+SA, RoT
TD-LTE: OFDM, MIMO, smart antenna

Standards:
TD-HSPA+: R8
TD-LTE: R8

Peak rate:
50 Mb/s (UL) and 100 Mb/s (DL)

Spectrum efficiency:
2–5 b/s/Hz

Under discussion

Fourth phase

TDD-based IMT-Advanced

Key techniques:
Dual topology (multihop and self-configuration), BR-OFDM, MIMO, smart antenna, cooperative communication, cognitive radio

Standard:
ITU-R

Peak rate:
100 Mb/s–1 Gb/s

Spectrum efficiency:
5–10 b/s/Hz

2000 2006 2007 2008 2009 2010 2015

Figure 8.1 Milestones of TD-SCDMA evolution. After Peng *et al.* (2010). ©2010 IEEE.

second phase. HSDPA and high-speed uplink packet access (HSUPA), known as HSPA (Holma *et al.* [2007]), apply adaptive modulation and coding (AMC), hybrid automatic repeat request (HARQ), fast packet scheduling (FPS), MIMO, and smart antenna technologies to increase their peak rates (Peng *et al.* [2010], Liu *et al.* [2006]).

In the downlink (DL) of LTE, orthogonal frequency division multiple access (OFDMA) is used and the subcarriers are allocated adaptively to the desired end-users to enhance multi-user diversity. More specifically, the scheduler of the base station, called enhanced NodeB (eNB), dynamically controls the time/frequency resources that are allocated to each end-user and adopts the multiplexing technique accordingly. In the uplink (UL) of LTE, both discrete Fourier transformation (DFT)-spread OFDM (SOFDM) and interleaved frequency division multiple access (IFDMA) are used. These single-carrier transmission techniques increase the UL inter-user orthogonality and enable the frequency-domain equalization efficiently (Peng *et al.* [2010]). In TD-SCDMA, different enhancement techniques are proposed to increase spectrum efficiency and peak rate of IMT-Advanced networks. For instance, the block repeat (BR)-OFDMA is considered to reduce the interferences of adjacent cells.

8.2 MAC layer

8.2.1 Resource allocation

The resource management and scheduling of LTE are monitored by the link layer, which consists of the following three sublayers (Larmo *et al.* [2009]): (*i*) packet data convergence protocol (PDCP) sublayer, which is responsible for handover, IP header compression, and ciphering; (*ii*) radio link control (RLC) sublayer, which performs data segmentation and concatenation; and (*iii*) medium access control (MAC) sublayer, which is responsible for resource management and scheduling. Compared with 3G cellular networks, the link layer protocols of LTE are designed to enhance quality-of-service (QoS) by increasing their reliability, security, and integrity.

In LTE networks, each end-user applies the scheduling request (SR) mechanism to send a request to access the channel. In the SR mechanism, an end-user reports the amount of data buffered in its queues, known as a buffer status report (BSR). Typically, the following two mechanisms are defined to perform resource management by means of SR: (*i*) dedicated SR (D-SR), where each end-user should send its SR on a dedicated resource over the physical uplink-control channel (PUCCH) to the eNB and (*ii*) random access-based SR (RA-SR), where the end-user should use a four-phase contention-based random access procedure (Larmo *et al.* [2009]). We note that RA-SR is used if no PUCCH resources for D-SR are assigned to the end-user.

Unlike conventional cellular networks, e.g., global system for mobile communications (GSM), where the channels are dedicated to high-priority QoS-enabled end-users, in LTE end-users apply a shared uplink and a shared downlink channel. In (Anas *et al.* [2008]), a combined admission control and time-frequency domain scheduling framework was proposed to support QoS and service differentiation for the LTE uplink. It is important to note that no specific scheduling and resource assignment scheme was standardized. The deployed scheduling algorithm in the eNB is responsible for meeting the QoS requirements of different traffic classes according to instantaneous channel conditions and/or fairness allocation policies (Kwan *et al.* [2009]). The scheduler selects both the appropriate end-user and radio bearer for downstream, while upstream, the uplink scheduling grants are assigned to end-users without specifying the radio bearer. Since the end-user sends BSRs for pre-allocated active radio bearers, the eNB ensures that users with high-priority data are given preference and obtain the assigned QoS. The scheduling grants are carried to end-users on the physical downlink-control channel (PDCCH) (Larmo *et al.* [2009]).

Different radio resource management algorithms have been proposed for LTE networks. In (Pedersen *et al.* [2009]), some of these algorithms, ranging from bearer admission control to semi-persistent, are compared. The simulation results demonstrate that each algorithm is suitable for a different type of service. For instance, it was shown that the semi-persistent scheduling algorithm outperforms the other resource management algorithms for highly loaded voice-over-IP (VoIP) end-users.

The scheduling and resource allocation algorithm for OFDM-based LTE networks was proposed in (Huang *et al.* [2009]). Using a dual decomposition approach, the convex optimization problem is defined and solved managing PHY layer resources (i.e., bandwidth and power) to maximize network utilization taking QoS support into account. Although the simulation results of the proposed algorithm demonstrates good performance for a single-cell LTE network, its complexity is a challenging issue, especially for multi-cell networks.

In (Luo *et al.* [2010]), a QoS-aware scheduling algorithm was introduced to enhance real-time video delivery over LTE networks using cross-layer techniques. In the proposed algorithm, various parameters are jointly considered in a cross-layer design framework (including network throughput, QoS requirements, and scheduling fairness) to dynamically allocate radio resources to multiple end-users. Experimental results show a significant performance enhancement of the proposed system. The results show that the network performance in terms of peak signal-to-noise ratio (PSNR) improves by using the cross-layer radio resource allocation algorithm. The proposed cross-layer scheduling algorithm dynamically changes the resource allocations based on the channel quality by optimizing the encoding parameters in order to ensure the highest received video quality under given delay constraints.

8.2.2 Retransmission

For deploying LTE under realistic wireless channel conditions, a two-layered retransmission scheme, called HARQ protocol, may be used to handle occasional retransmission errors. The MAC and RLC sublayers deploy the HARQ protocol to provide a highly reliable selective-repeat automatic repeat request (ARQ) protocol. This protocol aims at reliably transferring traffic with low latency and low overhead (Astely *et al.* [2009]).

Furthermore, LTE is able to perform channel-dependent scheduling in both time and frequency domains in order to synchronize itself with rapid channel-quality variations (i.e., fading) (Mongha *et al.* [2008]). For low data rate applications, e.g., VoIP, where the additional channel-dependent scheduling overhead is not efficient, LTE uses space-frequency block coding (SFBC) for transmit diversity (Astely *et al.* [2009]).

8.3 Power saving

Similar to other wireless broadband access networks, 3GPP defines the following three power saving modes: (i) active mode, (ii) discontinuous reception (DRX) in radio resource control (RRC) idle mode (RRC_{IDLE}), and (iii) DRX in connected sleep mode ($RRC_{CONNECTED}$). In the idle and sleep modes, mobile stations can turn off their transmitter/receiver equipment to decrease energy consumption. One of the important benefits of using power management mechanisms defined in 3GPP LTE is the small control signaling and network overhead. On the other hand, the proposed techniques provide more efficient power saving by using simpler operational procedures than existing power management techniques deployed in cellular networks (Kim and Mohanty [2010]).

Using DRX, LTE benefits from spatial reuse capacity techniques, such as over-the-air resource saving, in both UL and DL directions. One of the major enhancements in LTE networks is the connectivity of DRX-enabled end-users with a new eNB during their movement. Optimizing the DRX parameters is a challenging issue, which should be considered to enhance power saving without impacting on the network throughput and delay performance. In LTE DRX mode, an end-user is registered through the evolved packet system (EPS)-mobility management (EMM). The registered end-user can be paged for DL traffic, while it must request RRC connection for access to the channel to transmit its UL traffic (Bontu and Illidge [2009]). It was shown in (Bontu and Illidge [2009]) that the DRX mode parameters of the $RRC_{CONNECTED}$ mode play a key role in minimizing the network delay, especially under real-time delay-sensitive traffic. For instance, enabling DRX with a suitable DRX mode parameter results in about 40–45% power saving for a 10 frame/s video stream without significant impact on the quality of the video. For VoIP traffic, power savings of up to 60% can be achieved. In the RRC_{IDLE} state, the DRX cycle plays a key role in the performance of the network, which can be set based on the end-user's calling profile. Furthermore, the network re-entry time can be significantly improved by sending multiple paging messages to the end-user.

In (Kim *et al.* [2009]), an adjustable power management mechanism for the DRX framework was proposed, which uses hierarchical cascaded power gating (HCPG) and hierarchical multi-level clock gating (HMLCG). The HCPG is defined to maximize both static power reduction in the idle state and the dynamic power reduction in the transition between active and idle modes. The proposed HCPG consists of (i) eight different power domains for each function/module of the network and end-users to control and minimize the power consumption of each component separately; (ii) one always-on power domain. Using HCPG domains, the power consumption can be optimized by limiting the power supply to the minimal power domains according to each application scenario. In HMLCG, clock gating (CG) and frequency scaling (FS) techniques are used to avoid dynamic power dissipation. The reported experimental results show that the proposed adjustable power management minimizes the power saving of LTE networks significantly (Kim *et al.* [2009]).

8.4 Handover

Because LTE networks provide high-speed connectivity beside cellular network services, handovers play a key role in the mobility support of end-users. Although no specific handover mechanism is defined in the LTE standard, soft handover is not recommended to be used. Applying hard handover in LTE networks leads to various challenging issues, such as increased handover outage probability and delay. The semi-soft handover was proposed and the reported results show that using semi-soft handover not only attains the lowest handover outage probability but also improves the reliability of VoIP service over LTE networks (Han and Wu [2010]).

For each handover in an LTE network, the context of an end-user (i.e., end-user plane packets and control plane context) is relocated from its old eNB to the new eNB. This relocation has to be performed in a fast and seamless manner to guarantee the quality of the received services before and after handover (Zheng and Wigard [2008]). The problem of out-of-order packet delivery during handover was analyzed in (Racz *et al.* [2007]). In the proposed solution, fast packet forwarding from the source eNB to the target eNB is employed to ensure the correct delivery order of packets. The results show that the proposed relocation-based handover mechanism of LTE has no impact on the performance of services received by the end-user (Racz *et al.* [2007]).

8.5 LTE-Advanced

In 2008, LTE-Advanced (also known as requirements for further advancements for evolved-universal terrestrial radio access) was initiated to enhance LTE radio access in terms of system performance and capabilities (36.913 [2009]). In 3GPP TR 36.913 (36.913 [2009]), the requirement specifications of LTE-Advanced are defined. In an LTE-Advanced network, the peak data rate increases up to 1 Gb/s and 500 Mb/s for DL and UL streams, respectively. The spectrum efficiency of next-generation LTE networks is enhanced up to 30 b/s/Hz and 15 b/s/Hz in DL and UL directions, respectively. To support the requirements of international mobile telephony (IMT)-Advanced (ITU-R M.1645 [2004], ITU-R M.2134 [2008]), the following characteristics are considered for next-generation LTE networks (Astely *et al.* [2009]):

1. **Carrier aggregation**, where multiple fixed bandwidth carriers are aggregated to support huge transmission bandwidth with very high transmission rates, e.g., 20 MHz carriers are aggregated to a bandwidth of up to 100 MHz.
2. **Relaying support** to increase coverage and decrease deployment cost.

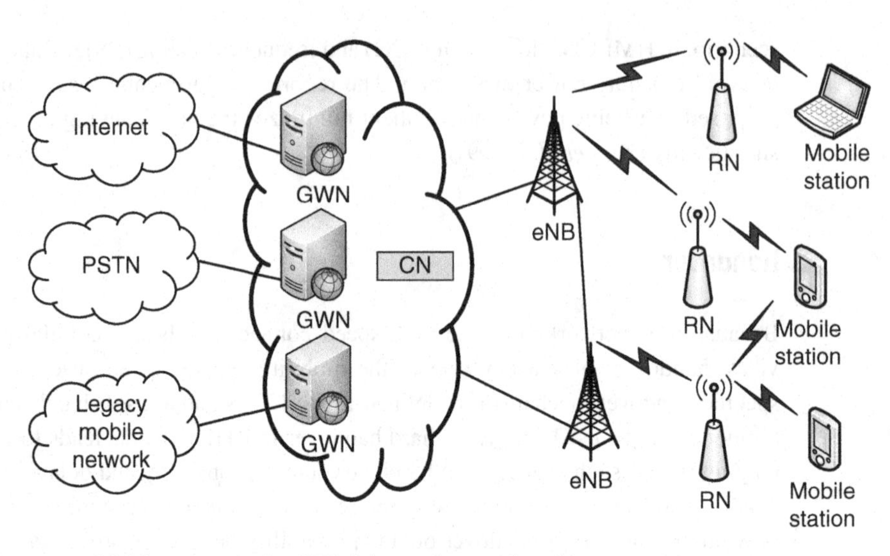

Figure 8.2 Multihop cooperative LTE-Advanced network architecture.

3. **Extended multi-antenna transmission** to increase the number of downlink/uplink transmissions, which enlarges the total transmission rate.
4. **Coordinated multipoint (CoMP) transmission/reception**, where multiple cells perform transmission/reception simultaneously to improve the performance of cell-edge nodes.
5. **Backward compatibility**, where both first LTE released end-users and LTE-Advanced end-users are able to access the network, thereby providing a cost-efficient pay-as-you-grow migration.

For relaying support in LTE-Advanced networks, the multihop cellular network architecture has been proposed (Zheng *et al.* [2009]). Figure 8.2 shows the architecture of a multihop cooperative LTE-Advanced network. In the multihop cooperative LTE-Advanced network architecture, the layer-2 functions of radio resource management are deployed in the remote nodes (RNs). More specifically, the RNs perform joint routing, link adaptation, and scheduling to help the eNB handle radio resources efficiently. The reported simulation results show that the multihop cooperative LTE-Advanced network outperforms single-hop networks in terms of signal-to-noise and interference ratio (SINR) and throughput (Zheng *et al.* [2009]).

9 Wireless mesh networks

Wireless mesh networks (WMNs) have been envisioned to enhance flexibility, increase reliability, and improve performance of wireless networks. Although WMNs are not widely considered in wireless metropolitan area network (WMAN) deployments, IEEE 802.11 wireless local area network (WLAN) is an interesting technology to realize low-cost WMNs. In this chapter, we provide an overview of the salient features and most important specifications of WiFi-based WMNs and describe the major challenging issues of routing and medium access control (MAC) protocols. Moreover, we briefly describe the optional mesh mode of initial standard IEEE 802.16d for fixed WiMAX, which has been removed from the IEEE standard 802.16e for mobile WiMAX.

9.1 Characteristics

Typically, there are two main approaches in the design of wireless networks (Schiller [2003]):

1. **Infrastructure networks:** Wireless mobile stations (STAs) rely on an underlying infrastructure for communication. They communicate with each other via a central control point, e.g., an access point (AP). WMANs, such as the global system for mobile communications (GSM) and universal mobile telecommunications system (UMTS), are typical examples for infrastructure wireless networks.
2. **Infrastructure-less networks:** STAs communicate directly with each other. In infrastructure-less wireless networks, also known as mobile ad-hoc networks (MANETs), STAs are able to act as routers. Emergency search-and-rescue operations, corporate meetings, and military communications in hostile terrains are example applications of MANETs.

The convergence of the two aforementioned wireless networking approaches leads to WMNs, which promise greater flexibility, increased reliability, and improved performance. WMNs employ multihop communications to forward traffic en route to and from a wired distribution system (DS), e.g., Internet (Akyildiz *et al.* [2005]). WMNs are expected to be widely deployed due to their ability to provide ubiquity, convenience, cost-efficiency, and simplicity (He *et al.* [2008]).

9.2 WiFi-based WMN

Although most deployed WLANs use the single-hop IEEE standard 802.11a/b/g or the emerging IEEE standard 802.11n, the recent amendment IEEE 802.11s provides mesh capability by converging the advantages of different routing protocols (Hiertz *et al.* [2008]). IEEE 802.11s defines the hybrid wireless mesh protocol (HWMP), which is inspired by a combination of (i) ad-hoc on-demand distance vector (AODV) routing for MAC address-based path selection and (ii) tree-based pro-active routing (Bahr [2009]). In addition to routing, WMNs face several challenges such as security and quality-of-service (QoS), which must be considered and addressed in IEEE 802.11s (Faccin *et al.* [2006]). Recently, IEEE 802.11s and its challenges have been discussed in greater detail in (Camp and Knightly [2008]) and some experimental work has been reported in (He *et al.* [2008], Ishmael *et al.* [2008], Garroppo *et al.* [2008]).

9.2.1 Routing protocols

During the last decade, numerous routing protocols have been developed for wireless networks. One common property of these protocols deals with the aspect of limitations inherent in the networks, i.e., limited bandwidth and high bit error rate (BER). Typically, wireless network routing protocols are categorized as follows:

1. **Table driven routing protocols:** Table driven, also known as pro-active, routing protocols attempt to maintain consistent, up-to-date routing information from each node to every other node in the wireless network. These protocols require that each participating node maintains one or more table in order to store the routing information. The nodes respond to changes in the network topology by propagating route updates throughout the network in order to maintain a consistent network view. The difference between table driven protocols lies in the number of routing tables they maintain as well as the methods used for maintaining the network connectivity.
2. **On-demand driven routing protocols:** On-demand driven, also known as re-active, routing protocols create routes only when desired by the source node (i.e., source-initiated). When a node requires a route to another node it initiates a *route discovery* process within the network and the process is completed when a route is found or all possible route permutations have been examined. Once a route has been established, it is maintained through some form of *route maintenance* procedure (e.g., cache-based) until either the destination node becomes inaccessible along every path from the source node or until the route is no longer desired.
3. **Hybrid routing protocols:** Hybrid routing protocols combine the advantages of both the aforementioned pro-active and re-active routing protocols, e.g., the zone routing protocol (ZRP) (Haas *et al.* [2002]). ZRP divides the network into different zones, which comprise the nodes in the transmission range of each other. Each node may be within multiple overlapping zones and each zone may have a different size. ZRP uses a pro-active routing protocol as intrazone routing protocol (IARP) and a reactive protocol as interzone routing protocol (IERP).

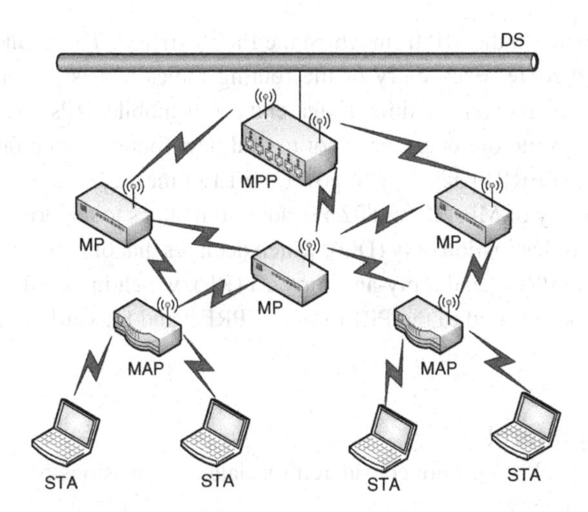

Figure 9.1 IEEE 802.11s wireless mesh network architecture.

Recently, IEEE 802.11s defined a novel hybrid routing protocol (i.e., HWMP), which applies AODV and tree-based routing protocols as re-active and pro-active routing protocols, respectively (Bahr [2007]). While AODV is used for peer-to-peer transmissions between fixed or mobile mesh nodes, called mesh points (MPs), the tree-based routing protocol is applied to provide MPs with access to the wired DS through fixed gateway nodes, called mesh portal points (MPPs) (Wang *et al.* [2008]). Figure 9.1 illustrates an example of IEEE 802.11s WMN architecture, where a mesh access point (MAP) is a special type of MP equipped with the additional capability of an AP to provide service to STAs. In the following, we describe AODV and the tree-based pro-active routing protocols proposed for HWMP in greater detail.

Re-active routing protocol

The AODV routing protocol, first presented in (Perkins and Belding-Royer [1999]) and then standardized in IETF RFC 3561 in 2003 (Perkins *et al.* [2003]), builds on the destination sequence distance vector (DSDV) algorithm (Perkins and Bhagwat [1994]). AODV improves the DSDV protocol by minimizing the number of broadcast packets. Similar to AODV, in IEEE 802.11s, when an MP desires to send a packet to another MP and does not already have a valid route to that, it initiates a *path discovery* process in order to locate the destination MP. The source MP broadcasts a path request (PREQ) packet to its neighbors, which in turn forward the request to their neighbors until either the destination MP or an intermediate MP with a route to the destination MP is located. It is worthwhile to note that AODV utilizes sequence numbers to ensure that all routes are loop-free and contain the most recent route information. During the process of forwarding the PREQ packets, intermediate MPs record the address of the transmitter neighbor in their routing tables for establishing a reverse shortest path. Once the first PREQ reaches the destination MP or an intermediate MP with a fresh enough route, a path reply (PREP) packet is sent back. As the PREP is routed back along the reverse path, the intermediate MPs along the forward route store the entries in their routing tables, which point to the MP from which the PREP arrived. These entries indicate the active forward route. Each entry of the routing tables consists of a lifetime parameter, which denotes the validity time of that entry. For mobile MPs, the protocol is able to re-initiate the route discovery protocol to find new routes for the failed path by means of path error (PERR) packets. Moreover, HELLO messages may be used to maintain local connectivity of MPs. IEEE 802.11s adds some flags to the original frame format of AODV such as destination only (DO), which declares that only the destination MP is able to respond to PREQ and reply-and-forward (RF), which in turn denotes that intermediate MPs should respond to the PREQ with a PREP and forward the PREQ to the destination MP.

Pro-active routing protocol

In IEEE 802.11s, two mechanisms are considered for data transmissions between MPs and MPPs (Wang *et al.* [2008]):

1. **Pro-active PREQ:** In this mechanism, MPPs periodically broadcast PREQ packets and the receiver MP creates/updates the path to the MPs by recording the metric and hop count parameters in their routing tables. Also, there is a pro-active PREP flag in the pro-active PREQ packet, which denotes that the receiver MP must send a PREP to the MPP immediately or in its next data packet.
2. **Pro-active RANN:** In this method, MPPs periodically broadcast root announcement (RANN) packets, which denote the availability and location of MPPs in the network. Each receiving MP creates/updates a route to the MPPs in its routing table and responds with a PREQ. Once the MPP receives the PREQ, it replies with a PREP to the MP.

In both above-mentioned mechanisms, MPs select the MPP with the smallest number of hops in case there is more than one MPP in the WMN. Although IEEE 802.11s does not specify which of these pro-active routing protocol mechanisms should be used, only one of them is allowed to be applied at any given time.

Recently, different techniques have been considered to improve the performance of AODV. In (Guezouri and Ouamri [2007]), three different optimization techniques are described for AODV: (i) reverse path setup – shortest path with minimum number of hops is selected; (ii) forward path setup – each intermediate node, which receives a PREQ and has a shortest path to the destination node informs the source node by sending a PREP packet, and (iii) route scattering – during the route discovery process, the source and destination nodes learn the routes to the in-path intermediate nodes. Recently, a novel method has been proposed to control the congestion on the selected path of the AODV routing protocol via an additional delay in broadcasting PREQ packets inserted by intermediate nodes (Xia *et al.* [2009]). The use of the global positioning system (GPS) is another approach to improving the performance of AODV in terms of overhead by reducing the PREQ broadcasting area (Espes and Mammeri [2007]). GPS may also be used for finding a reliable path by monitoring the directions of mobile nodes in MANETs (Qiang and Hongbo [2008]).

Since in both WMNs and MANETs intermediate nodes relay incoming packets to the destination node, the collaboration between intermediate nodes and source/destination nodes has been studied in greater detail in (Demir and Comaniciu [2007]). In this study, a novel auction-based technique is proposed to motivate selfish nodes (i.e., nodes which do not cooperate with other nodes in the routing process) to collaborate with other nodes. The reported results show that by using this technique a MANET with fully selfish nodes performs in a similar way to an unselfish MANET.

IEEE 802.11s introduces a radio metric-AODV (RM-AODV) as a default path selection approach using the airtime link metric (Bahr [2006]). This parameter denotes the amount of consumed channel resources when transmitting a prespecified test frame over a link. The expected transmission count (ETX) metric is another metric proposed for WMNs (He *et al.* [2008]). ETX applies the successful forward delivery ratio and successful reverse delivery ratio to select the path with the highest packet delivery ratio. In (Draves *et al.* [2004]), the expected transmission time (ETT) is proposed to assign weights to individual links based on their packet loss rate and bandwidth. Weighted cumulative ETT (WCETT), which accounts for the interference between links using the same frequency channel, is proposed and deployed together with ETT. The reported results show an improved throughput performance of the ETX metric for a multi-radio multi-channel WMN. The metric of interference and channel (MIC) switching technique is proposed in (Yang *et al.* [2005]). The MIC considers the set of the transmitter node's neighbors, which are affected by the transmission and number of channel switching. The interference aware routing metric, which applies the interference ratio of each channel in its path selection algorithm, is proposed in (Subramanian *et al.* [2006]). The proposed path selection algorithm improves the AODV network throughput in comparison with WCETT.

In (Katti *et al.* [2008]), a novel WMN architecture, called COPE, is proposed, where intermediate MPs aggregate in-transit packets using XOR coding techniques in order to improve network throughput. A heuristic load-balanced coding-aware routing (HLCR) mechanism and a novel path metric called heuristic path metric for coding and load-balancing (HPMCL) is proposed in (Fan *et al.* [2009]). HPMCL selects the path based on network loads, expected number of transmissions, and coding opportunities. The reported results show that HLCR improves the network throughput in comparison with COPE.

9.2.2 MAC protocols

The IEEE standard 802.11s introduces an optional contention-less access mechanism, called *mesh deterministic access (MDA)*, to provide end-to-end QoS for delay-sensitive traffic (Hiertz *et al.* [2008]). MDA is a scheduling-based medium reservation scheme. In MDA, MPs reserve the wireless medium for MDA opportunities (MDAOPs) using the following two types of time periods: (*i*) *neighborhood MDAOP times* – these time periods denote the transmitting/receiving (TX/RX) times of a given MP during which the MP and its neighboring MPs are either a transmitter or receiver of the corresponding MDAOPs, and (*ii*) *MDAOP interfering times* – these time periods keep track of the

MDAOPs of neighboring MPs, which the MP is aware of and during which it is not involved as a transmitter or receiver. A new MDAOP can be set up between a pair of MPs if there is no overlap of both MPs' neighborhood MDAOP and MDAOP interfering times (Wang *et al.* [2008]). Moreover, a QoS-driven MAC-layer resource allocation scheme was proposed for WMNs in (Cheng and Zhuang [2009]).

9.3 WiMAX-based WMN

9.3.1 Architecture

In a fixed WiMAX-based WMN, an SS is called mesh SS (MSS) and the BS is called mesh BS (MBS). Unlike in the point-to-multipoint mode, where the communications among SSs are done through the BS, in the mesh mode transmissions use the direct links between MSSs. Upon initialization, each MSS establishes one link with each node in its range. Typically, each node has a parent node, which is the node with the fewest number of hops to the MBS and has the highest signal-to-noise ratio (SNR) among its neighbors. The links between MSSs and their parents form a scheduling tree. However, this scheduling tree is not guaranteed to be optimal (Kuran and Tugcu [2007]).

9.3.2 Scheduling

Similar to the point-to-multipoint mode, at the MAC layer of WiMAX-based WMNs the requested services of each SS are first registered during the initialization phase and subsequently the connections are established.

To implement WMNs, IEEE 802.16d fixed WiMAX applies two scheduling types: (i) centralized and (ii) distributed. In the centralized scheduling mode, each MSS sends its request to the MBS that manages the network. The MBS determines the flow assignment from the resource request of MSSs. In the distributed scheduling mode, each MSS distributes its scheduling information and one-hop neighbors among all its adjacent MSSs. Therefore, every MSS has the scheduling information of all its two-hop neighbors and makes its scheduling decisions based on this information. The distributed scheduling is further subdivided into coordinated and uncoordinated methods. In the coordinated method, the scheduling information is sent in a collision-free manner, whereas in the uncoordinated method collisions are possible. It is worthwhile to mention that the two different mesh scheduling methods can be applied together by subdividing the data part of the frame into two parts, one for centralized scheduling and another one for distributed scheduling (Kuran and Tugcu [2007]). Note that no specific scheduling algorithm is standardized to support different types of data flows and QoS for the MSSs of fixed WiMAX-based WMNs, but various scheduling algorithms can be found in (Abu Ali *et al.* [2008]).

Part IV

FiWi access networks

Traditionally, wireless and optical fiber networks have been designed separately from each other. Wireless networks are aimed at meeting specific service requirements while coping with particular transmission impairments and optimizing the utilization of the system resources to ensure cost-effectiveness and satisfaction for the user. In optical networks, on the other hand, research efforts rather focused on cost reduction, simplicity, and future-proofness against legacy and emerging services and applications by means of optical transparency. Wireless and optical access networks can be thought of as complementary. Optical fiber does not go everywhere, but where it does go, it provides a huge amount of available bandwidth. Wireless access networks, on the other hand, potentially go almost everywhere, but provide a highly bandwidth-constrained transmission channel susceptible to a variety of impairments.

Future broadband access networks not only have to provide access to information when we need it, where we need it, and in whatever format we need it, but also, and arguably more importantly, have to bridge the digital divide and offer simplicity and user-friendliness based on open standards in order to stimulate the design of new applications and services. Toward this end, future broadband access networks must leverage on both optical and wireless technologies and converge them seamlessly, giving rise to *fiber-wireless (FiWi)* access networks (Aissa and Maier [2007]). FiWi access networks are instrumental in strengthening our information society while avoiding its digital divide. By combining the capacity of optical fiber networks with the ubiquity and mobility of wireless networks, FiWi networks form a powerful platform for the support and creation of emerging as well as future unforeseen applications and services, e.g., telepresence. FiWi networks hold great promise to change the way we live and work by replacing commuting with teleworking. This not only provides more time for professional and personal activities for corporate and our own personal benefit, but also helps reduce fuel consumption and protect the environment, issues that are becoming increasingly important in our lives.

Due to the difficulty and prohibitive costs of supplying optical fiber to all end-user premises as well as the spectrum limitations of wireless access networks, bimodal FiWi access networks are more attractive than relying on either stand-alone access solution. FiWi networks are realized by integrating wireless access technologies, e.g., cellular, WiMAX, and WiFi, with installed optical fiber infrastructure that has been pushed ever closer toward end-users over the last few years and has been the preferred medium of choice in most of today's greenfield deployments, where fiber rather than copper cables are installed for broadband access. To better understand the rationale behind the vision of FiWi access networks, we first describe related research topics and then define FiWi networks as a new research area in the following:

- **Fixed mobile convergence (FMC)**
  According to the European Telecommunications Standardization Institute (ETSI), FMC is concerned with developing network capabilities and supporting standards that may be used to seamlessly offer a set of consistent services via fixed or mobile access to fixed or mobile, public or private, networks, independently of the access technique (MacLeod and Safavian [2008]).

FMC can be achieved at different levels, e.g., business or service provisioning level. Note, however, that FMC does not necessarily imply the physical convergence of networks. In fact, the convergence at the network facilities level, where an operator uses the same physical network infrastructure with common transmission and switching systems to provide both mobile and fixed services, is more accurately referred to as *fixed mobile integration (FMI)* (Harrison and Hearnden [1999]). While FMI considers any type of access networks, e.g., digital subscriber line (DSL), hybrid fiber-coax (HFC), or wireless local area network (WLAN), the so-called *optical wireless integration (OWI)* focuses on emerging optical and wireless broadband access technologies, as explained next.

- **Optical wireless integration (OWI)**
  Current copper-based access network technologies such as DSL and HFC face serious challenges to meet the requirements of future broadband access networks. While DSL suffers from severe distance and noise limitations, HFC is not able to carry data traffic efficiently due to its upstream noise and crosstalk accumulation. Recent progress in optical fiber technologies, especially the maturity of integration and new packaging technologies, has rendered optical fiber access networks a promising low-cost broadband solution. In particular, passive optical networks (PONs) are able to provide lower network deployment and maintenance costs as well as longer distances than current DSL and HFC networks. OWI aims at integrating PONs and other optical fiber access technologies with emerging broadband wireless access technologies, e.g., WiMAX, in order to increase the capacity of wireless access networks and reduce access point complexity through centralized management (Luo *et al.* [2006]). It is important to note that there is a difference between OWI and free-space optical wireless (OW) communications. OW systems to test line-of-sight wireless communications were already designed in the 1960s, well before the development of optical Þber communications (Davis *et al.* [2003]). Current radio frequency (RF) wireless systems, e.g., WiFi, WiMAX, or ultra-wideband (UWB), are limited in both data rate and range due to their low carrier frequencies. OW communications links operate at much higher carrier frequencies than their RF counterparts. As the RF spectrum becomes increasingly congested, OW represents an interesting alternative to RF wireless systems. OW may be deployed as a temporary backbone for rapidly deployable mobile wireless communication infrastructure, especially in densely populated urban areas. Note, however, that unlike OWI networks, OW links and networks do not involve any wired (fiber) infrastructure.

- **Radio-over-fiber (RoF) vs. radio-and-fiber (R&F) networks**
  RoF networks have been studied for many years as an approach to integrate optical fiber and wireless networks. In RoF networks, RFs are carried over optical fiber links between a central office (CO) and multiple low-cost remote antenna units (RAUs) in support of a variety of wireless applications, e.g., microcellular radio systems (Chu and Gans [1991]). It was experimentally demonstrated that RoF

networks can have an optical fiber range of up to 50 km. However, inserting an optical distribution system in wireless networks may have a major impact on the performance of medium access control (MAC) protocols (Dang and Niemegeers [2005]), as discussed in greater detail in Chapter 10. The limitations of RoF networks can be avoided in so-called *radio-and-fiber (R&F) networks* (Henry [2007]). While RoF networks use optical fiber as an analog transmission medium between a CO and one or more RAUs with the CO being in charge of controlling access to both optical and wireless media, in R&F networks access to the optical and wireless media is controlled separately from each other by using in general two different MAC protocols in the optical and wireless media, with protocol translation taking place at their interface. As a consequence, wireless MAC frames do not have to travel along the optical fiber to be processed at the CO, but simply traverse their associated AP and remain in the wireless front-end, thus avoiding the negative impact of fiber propagation delay on the network performance.

- **FiWi access networks**
FiWi networks may deploy both RoF and R&F technologies. By simultaneously providing wired and wireless services over the same infrastructure, FiWi networks are able to consolidate optical and wireless access networks that are usually run independently of each other, thus potentially leading to major cost savings.
FiWi networking research deals with the OWI of emerging optical and wireless broadband access technologies, e.g., wireless mesh network (WMN). FiWi research focuses on the physical (PHY), MAC, and network layers with the goal of developing and investigate low-cost enabling FiWi technologies as well as layer-2 and layer-3 protocols and algorithms. Higher-layer network capabilities developed through FMC standardization efforts can be exploited on top of the PHY, MAC, and network layers of FiWi networks. As we will see in the subsequent chapters, FiWi research investigates new methods of optical RF generation exploiting fiber nonlinearities and various modulation techniques. It also includes the study of different remodulation schemes for the design of colorless (i.e., wavelength-independent) RAUs. While significant progress has been made at the PHY layer of FiWi and in particular RoF transmission systems, FiWi networking research on layer-2 and layer-3 related issues has begun only very recently. Among others, FiWi layer-2/3 research includes the joint optimization of performance-enhancing MAC mechanisms separately used in the wireless and optical network segments, e.g., wireless frame aggregation and optical burst assembly, hybrid access control protocols, integrated path selection algorithms, as well as advanced resilience techniques. Layer-2/3 networking research is crucial to unleash the full potential of FiWi networks (Sarkar *et al.* [2007a], Ghazisaidi and Maier [2011]).

In the following, we discuss the aforementioned challenges and opportunities of FiWi access networks in technically greater detail. The remainder of this part is structured as follows. In Chapter 10, we elaborate on the difference between *RoF and R&F networks* and summarize some of the key enabling RoF and R&F technologies. In addition, we highlight two recently proposed RoF and R&F based state-of-the-art FiWi network testbeds and describe lessons learned. Furthermore, we outline the challenges and open issues of layer-2/3 protocols and algorithms in future FiWi access networks. Chapter 11 provides a more comprehensive overview of possible FiWi network *architectures* based on either RoF or R&F technologies. The *planning and reconfiguration* of various FiWi networks, including optimal ONU placement and direct inter-ONU communications, are discussed in Chapter 12. In Chapter 13, we perform a *techno-economic analysis* of Ethernet PON (EPON) and WiMAX, which play a key role in many previously studied FiWi broadband access networks. The opportunities of powerful *network coding* techniques to enhance the throughput-delay performance of next-generation PONs (NG-PONs) and FiWi access networks are explored in Chapter 14. In Chapter 15, we introduce and analyze a variety of *optical and wireless protection* techniques to improve the survivability of NG-PONs and FiWi networks and render them robust against multiple fiber link failures. Chapter 16 explores *hierarchical frame aggregation* techniques for FiWi networks based on integrated EPON and WiFi mesh networks. The presented simulation and experimental results show that the proposed techniques help improve the throughput-delay performance, jitter, packet loss, and mean opinion score of FiWi networks significantly. The wireless mesh front-end of FiWi access networks provides multiple possible paths to route traffic to any ONU of the optical backhaul PON. In Chapter 17, we review state-of-the-art *wireless and integrated routing algorithms* for FiWi access networks, including recently proposed energy-aware routing for "green" energy-efficient broadband access networks. The chapter also elaborates on techniques to guarantee *QoS continuity* across the optical–wireless interface of FiWi access networks. Finally, Chapter 18 outlines interesting opportunities as to how FiWi broadband access networks can be also adopted in other relevant sectors, e.g., power grids, and be used to realize the communications infrastructure of the future *smart grid.*

10 RoF vs. R&F networks

Radio-over-fiber (RoF) networks have been studied for many years as an approach to integrating optical fiber and wireless networks. In RoF networks, radio frequencies (RFs) are carried over optical fiber links between a central station and multiple low-cost remote antenna units (RAUs) in support of a variety of wireless applications. For instance, a distributed antenna system connected to the base station of a microcellular radio system via optical fibers was proposed in (Chu and Gans [1991]). To efficiently support time-varying traffic between the central station and its attached base stations, a centralized dynamic channel assignment method is applied at the central station of the proposed fiber optic microcellular radio system. To avoid having to equip each radio port in a fiber optic microcellular radio network with a laser and its associated circuit to control the laser parameters such as temperature, output power, and linearity, a cost-effective radio port architecture deploying remote modulation may be used (Wu *et al.* [1994]).

Apart from realizing low-cost microcellular radio networks, optical fibers can also be used to support a wide variety of other radio signals. RoF networks are attractive since they provide transparency against modulation techniques and are able to support various digital formats and wireless standards in a cost-effective manner. It was experimentally demonstrated in (Tang *et al.* [2004]) that RoF networks are well suited to simultaneously transmit wideband code division multiple access (WCDMA), IEEE 802.11a/g wireless local area network (WLAN), personal handyphone system (PHS), and global system for mobile communications (GSM) signals. Figure 10.1 illustrates the method investigated in (Tang *et al.* [2004]) for two different radio client signals transmitted by the central station on a single-mode fiber (SMF) downlink to a base station and onward to a mobile user or vehicle. At the central station, both radio client signals are first up-converted to a higher frequency by using a frequency converter. Then the two RF signals go into two different electroabsorption modulators (EAMs) and modulate the optical carrier wavelength emitted by two separate laser diodes. An optical combiner combines the two optical signals onto the SMF downlink. At the base station, a photodiode converts the incoming optical signal to the electrical domain and radiates the amplified signal through an antenna to a mobile user or vehicle which uses two separate frequency converters to retrieve the two different radio client signals.

While SMFs are typically found in outdoor optical networks, many buildings have preinstalled multimode fiber (MMF) cables. Cost-effective multimode fiber (MMF)-based networks can be realized by deploying low-cost vertical cavity surface emitting

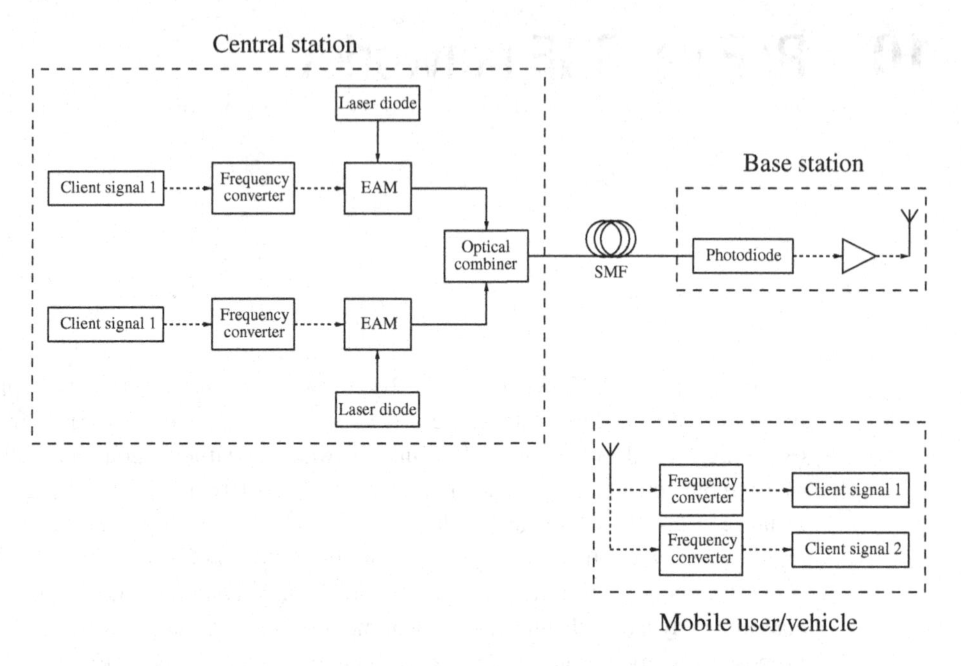

Figure 10.1 Radio-over-SMF network downlink using EAMs for different radio client signals. After Tang *et al.* (2004). ©2004 IEEE.

lasers (VCSELs). In (Lethien *et al.* [2005]), different kinds of MMF in conjunction with commercial off-the-shelf (COTS) components were experimentally tested to demonstrate the feasibility of indoor radio-over-MMF networks for the in-building coverage of second-generation (GSM) and third-generation cellular radio networks (universal mobile telecommunications system [UMTS]) as well as IEEE 802.11a/b/g WLAN and digital enhanced cordless telecommunication packet radio service (DECT PRS).

To realize future multiservice access networks, it is important to integrate RoF systems with existing optical access networks. In (Lin *et al.* [2007a]), a novel approach for simultaneous modulation and transmission of both RoF RF and FTTH baseband signals using a single external integrated modulator was experimentally demonstrated, as shown in Fig. 10.2. The external integrated modulator consists of three different Mach-Zehnder modulators (MZMs) 1, 2, and 3. MZM 1 and MZM 2 are embedded in the two arms of MZM 3. The RoF RF and FTTH baseband signals independently modulate the optical carrier generated by a common laser diode by using MZM 1 and MZM 2, respectively. Subsequently, the optical wireless RF and wired-line baseband signals are combined at MZM 3. After propagation over an SMF downlink, an optical filter (e.g., fiber grating) is used to separate the two signals and forward them to the wireless and FTTH application, respectively. It was experimentally demonstrated that a 1.25 Gb/s baseband signal and a 20-GHz 622 Mb/s RF signal can be simultaneously modulated and transmitted over 50 km standard SMF with acceptable performance penalties.

The aforementioned research projects successfully demonstrated the feasibility and maturity of low-cost multiservice RoF networks. Their focus was on the investigation

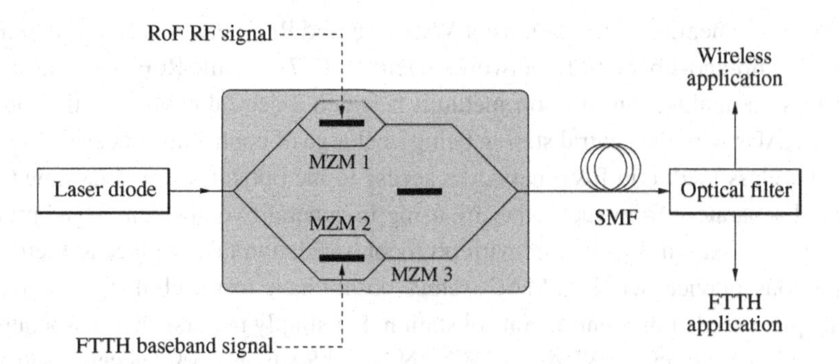

Figure 10.2 Simultaneous modulation and transmission of FTTH baseband signal and RoF RF signal using an external integrated modulator consisting of three Mach-Zehnder modulators (MZMs). After Lin *et al.* (2007a). ©2007 IEEE.

of RoF transmission characteristics and modulation techniques, considering primarily physical layer related performance metrics, e.g., power penalty, error vector magnitude (EVM), and bit error rate (BER) measurements. It was shown that RoF networks can have an optical fiber range of up to 50 km. However, inserting an optical distribution system in wireless networks may have a major impact on the performance of medium access control (MAC) protocols (Dang and Niemegeers [2005]). The additional propagation delay may exceed certain timeouts of wireless MAC protocols, resulting in a deteriorated network performance. More precisely, MAC protocols based on centralized polling and scheduling, e.g., IEEE 802.16 WiMAX, are less affected by increased propagation delays due to their ability to take longer walk times between central station and wireless subscriber stations into account by means of interleaved polling and scheduling of upstream transmissions originating from different subscriber stations. However, in distributed MAC protocols, e.g., the widely deployed distributed coordination function (DCF) in IEEE 802.11a/b/g WLANs, the additional propagation delay between wireless stations and access point poses severe challenges. To see this, note that in WLANs a source station starts a timer after each frame transmission and waits for the acknowledgment (ACK) from the destination station. By default the ACK timeout value is set to 9 μs and 20 μs in 802.11a/g and 802.11b WLAN networks, respectively. If the source station does not receive the ACK before the ACK timeout it will resend the frame for a certain number of retransmission attempts. Clearly, one solution to compensate for the additional fiber propagation delay is to increase the ACK timeout. Note, however, that in DCF the ACK timeout must not exceed the DCF interframe space (DIFS), which prevents other stations from accessing the wireless medium and thus avoiding collision with the ACK frame (in IEEE 802.11 WLAN specifications DIFS is set to 50 μs). Due to the ACK timeout, optical fiber can be deployed in WLAN-based RoF networks only up to a maximum length. For instance, it was shown in (Kalantarisabet and Mitchell [2006]) that in a standard 802.11b WLAN network the fiber length must be less than 1948 m to ensure the proper operation of DCF. In addition, it was shown that there is a trade-off between fiber length and network throughput. As more fiber is deployed the network throughput decreases gradually.

The aforementioned limitations of WLAN-based RoF networks can be avoided in so-called radio-and-fiber (R&F) networks (Henry [2007]). While RoF networks use optical fiber as an analog transmission medium between a central control station and one or more RAUs with the central station being in charge of controlling access to both optical and wireless media, in R&F networks access to the optical and wireless media is controlled separately from each other by using in general two different MAC protocols in the optical and wireless media, with protocol translation taking place at their interface. As a consequence, wireless MAC frames do not have to travel along the optical fiber to be processed at the central control station, but simply traverse their associated access point and remain in the WLAN. In WLAN-based R&F networks, access control is done locally inside the WLAN without involving any central control station, thus avoiding the negative impact of fiber propagation delay on the network throughput.

R&F networks are well suited to build WLAN-based FiWi networks of extended coverage without imposing stringent limits on the size of the optical backhaul, as opposed to RoF networks that limit the length of deployed fibers to a couple of kilometers. Recall that this holds only for distributed MAC protocols such as DCF, but not for MAC protocols that deploy centralized polling and scheduling, e.g., WiMAX and cellular networks (Kim *et al.* [2010], Gong *et al.* [2010]). The pros and cons of different possible broadband access architectures to integrate WiMAX with Ethernet passive optical network (EPON) technologies, including unified connection-oriented and microwave-over-fiber (MoF) architectures, were discussed in detail in (Shen *et al.* [2007]).

10.1 Enabling technologies

Both RoF and R&F technologies can be found in FiWi access networks. In this section, we discuss enabling technologies in greater detail.

10.1.1 RoF technologies

Several RoF technologies have been emerging for the realization of low-cost FiWi access networks. In the following, we briefly summarize some of the key enabling RoF technologies. For further details and a technically more profound discussion, we refer the interested reader to (Jia *et al.* [2007]).

Optical RF generation

To avoid the electronic bottleneck, the generation of RF signals is best done optically. The following novel optical RF generation techniques were experimentally studied and demonstrated in (Jia *et al.* [2007]):

- **FWM in HNL-DSF:** Four-wave mixing (FWM) in a highly nonlinear dispersion-shifted fiber (HNL-DSF) can be used to realize simultaneous all-optical up-conversion of multiple wavelength channels by using optical carrier suppression (OCS) techniques. FWM is transparent to the bit rate and modulation format, which may be different on each wavelength. Due to the ultrafast response of HNL-DSF, Terahertz optical RF generation is possible.

- **XPM in HNL-DSF:** Cross-phase modulation (XPM) in a nonlinear optical loop mirror (NOLM) in conjunction with straight pass in HNL-DSF enables the all-optical up-conversion of multiple wavelength channels without any interference- and saturation-effect limitation.
- **XAM in EAM:** All-optical wavelength up-conversion by means of cross-absorption modulation (XAM) in an EAM has several advantages such as low power consumption, compact size, polarization insensitivity, and easy integration with other devices.
- **External IM:** External intensity modulation (IM) is another approach for optical RF generation, deploying one of the three following modulation schemes: double-sideband (DSB), single-sideband (SSB), and OCS.
- **External PM:** Instead of external IM, external phase modulation (PM) can be used for optical RF generation.

According to Jia *et al.* [2007], external intensity and phase modulation schemes are the most practical solutions for all-optical RF generation due to their low cost, simplicity, and long-distance transmission performance.

Remote modulation

An interesting approach to building low-cost FiWi networks is the use of a single light source at the central office (CO) to generate a downlink wavelength that is reused at RAUs for upstream transmission by means of remote modulation, thereby avoiding the need for an additional light source at each RAU. The following remodulation schemes were experimentally studied in (Jia *et al.* [2007]):

- **DPSK for downstream/OOK for upstream:** PM is deployed to generate a differential phase-shift-keyed (DPSK) optical downstream signal. The DPSK is up-converted through OCS modulation. An optical splitter is used at each RAU to divide the arriving optical signal into two parts. One part is demodulated by a Mach-Zehnder interferometer and is subsequently detected by a photodetector. The other part is on–off-keyed (OOK) remodulated with upstream data using an MZM and is sent to the CO.
- **OCS for downstream/reuse for upstream:** At the CO, an optical carrier is split prior to optical RF generation by means of OCS and is then combined with the RF signal and sent downstream. Each RAU utilizes a fiber Bragg grating (FBG) to reflect the optical carrier while letting the RF signal pass to a photodetector. The reflected optical carrier is remodulated with upstream data and is then sent back to the CO.
- **PM for downstream/directly modulated SOA for upstream:** Similar to the aforementioned scheme, an optical carrier is combined with an RF signal, generated by means of PM, and sent downstream where an FBG is used at the RAU to reflect the optical carrier and pass the RF signal. The reflected optical carrier is amplified and directly modulated with upstream data using a semiconductor optical amplifier (SOA).

The use of a colorless (i.e., wavelength-independent) SOA as an amplifier and modulator for upstream transmission provides a promising low-cost RoF solution that is easy to maintain (Jia *et al.* [2007]).

10.1.2 R&F technologies

R&F-based FiWi access networks may deploy a number of enabling optical and wireless technologies.

Optical technologies

Apart from passive optical networks (PONs), the following optical technologies are expected to play an increasingly important role in the design of a flexible and cost-effective optical backhaul for FiWi networks (Kazovsky *et al.* [2007]):

- **Tunable lasers:** Directly modulated external cavity lasers, multisection distributed feedback (DFB)/distributed Bragg reflector (DBR) lasers, and tunable VCSELs can be used as tunable lasers that render the network flexible and reconfigurable and help minimize production cost and reduce backup stock.
- **Tunable receivers:** A tunable receiver can be realized by using a tunable optical filter and a broadband photodiode. Other more involved implementations exist (see Kazovsky *et al.* [2007]).
- **Colorless ONUs:** Reflective SOAs (RSOAs) can be used to build colorless optical network units (ONUs) that remotely modulate optical signals generated by centralized light sources.
- **Burst-mode laser drivers:** Burst-mode transmitters are required for ONUs. They have to be equipped with laser drivers that provide fast burst on/off speed, sufficient power suppression during idle period, and stable, accurate power emission during burst transmission.
- **Burst-mode receivers:** Burst-mode receivers are required at the central optical line terminal (OLT) of a PON and must exhibit a high sensitivity, wide dynamic range, and fast time response to arriving bursts. Among others, design challenges for burst-mode receivers include dynamic sensitivity recovery, fast level recovery, and fast clock recovery.

Wireless technologies

A plethora of broadband wireless access technologies exist. For a comprehensive survey on available broadband wireless access technologies we refer the interested reader to (Kuran and Tugcu [2007]). Currently, the three most important ones for the implementation of the wireless part of FiWi networks are WiFi, WiMAX, and wireless mesh network (WMN), which were described at length in Chapters 6, 7, and 9, respectively.

10.2 State-of-the-art testbeds

In this section, we briefly describe two recently proposed FiWi network testbeds and highlight the lessons learned.

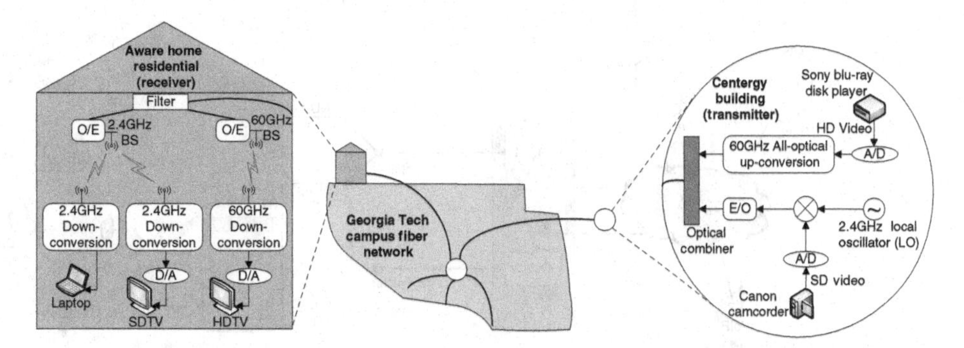

Figure 10.3 Georgia Institute of Technology RoF field demonstration of SD/HD video delivery using 2.4 GHz and 60 GHz millimeter-wave transmissions. After Chowdhury *et al.* (2009a). ©2009 IEEE.

10.2.1 RoF testbed

Figure 10.3 shows the RoF testbed designed at the Georgia Institute of Technology for the field trial demonstration of 270 Mb/s standard definition (SD) and 1.485 Gb/s high definition (HD) real-time video stream delivery using 2.4 GHz and 60 GHz millimeter (mm)-wave transmissions over 2.5 km SMF between the Centergy building (transmitter) and the Aware home residential building (receiver) (Chowdhury *et al.* [2009a]). All-optical up-conversion is used at the transmitter to generate a 60 GHz mm-wave signal (by means of phase modulation (PM)) and to send the HD video signal at 1554 nm. As shown in Fig. 10.3, electrical mixing and double-sideband optical modulation techniques are used to up-convert the SD video 2.4 GHz radio signal before optical transmission at 1550 nm. PIN photodiodes are used at the receiver to perform optical–electrical (OE) conversion of the filtered optical signals. The experimental results demonstrate a very good BER performance of the received video signals.

10.2.2 R&F testbed

Figure 10.4 shows the University of California (UC) Davis R&F testbed integration of two Ethernet PONs (EPONs) and an IEEE 802.11g WLAN-based WMN with a maximum transmission rate of 54 Mb/s for voice, video, and data traffic (Chowdhury *et al.* [2009a]). In this architecture, optical protection is provided by using full PON duplication. Programmability was realized by using a separate Linux PC connected to each ONU and open source firmware in each wireless gateway and router. The results show that the quality of video transmissions sharply deteriorates for an increasing number of wireless hops. In fact, the video client showed a blank screen after four wireless hops. The experimental results clearly show that running EPON and WMN networks independently yields a poor FiWi network performance. A more involved experimental investigation of integrated FiWi network architectures is needed taking hybrid MAC protocols, integrated path selection algorithms, and advanced resilience techniques into account.

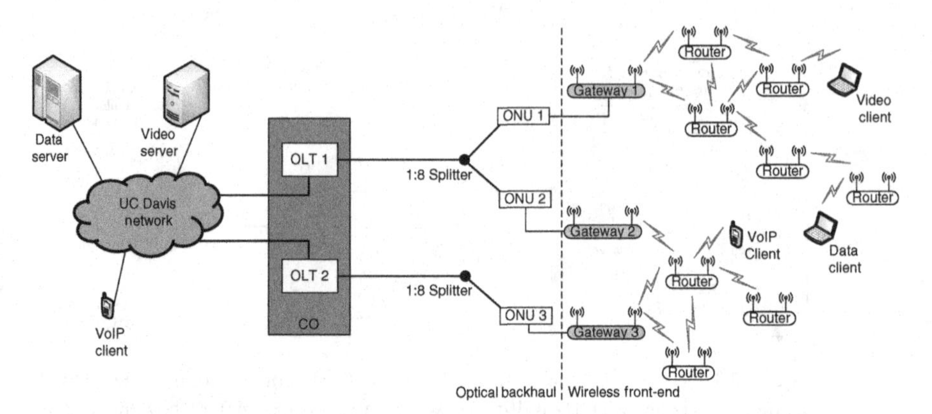

Figure 10.4 UC Davis R&F testbed integration of EPON and WMN for voice, video, and data traffic. After Chowdhury *et al.* (2009b). ©2009 IEEE.

10.3 Challenges and open issues

Most RoF related work has focused on the physical (PHY) layer, with a particular focus on downstream transmissions, in an effort to mitigate the detrimental impact of various transmission impairments, e.g., low opto-electronic conversion efficiency, fiber chromatic dispersion, and degradation due to fiber nonlinearities (Lim *et al.* [2010]). There exist a number of different physical layer networking options, including RoF, intermediate frequency (IF) over fiber, baseband over fiber, and recently reported *digitized RoF* (Lim and Nirmalathas [2010]). Conventional RoF networks face a number of challenges. RoF networks require special techniques for mitigating the effects of intermodulation distortion arising from the nonlinear device characteristics of optical components. Furthermore, RoF networks often require opto-electronic devices with wide bandwidth that need to be customized to meet specific application demands. As a result, RoF networks can be less competitive than wireless alternatives for the backhaul. To mitigate these shortcomings, the transmission of digitized RF signals may be a more cost-effective alternative to conventional analog RoF links by exploiting the better performance of digital optical links and the benefits of direct RF distribution to realize simpler base stations (Nirmalathas *et al.* [2009, 2010]). Several cost-effective technologies for very-high-throughput RoF networks were recently investigated and demonstrated, including multicarrier orthogonal frequency division multiplexing (OFDM) and single-carrier multilevel modulation (Yu *et al.* [2010]). Advanced technologies, e.g., polarization diversity RoF, for in-building RoF networks were proposed and experimentally investigated in (Xu *et al.* [2010]).

While significant progress has been made at the PHY layer of FiWi and in particular RoF transmission systems, FiWi networking research on layer-2/3 related issues has begun only very recently (Ghazisaidi and Maier [2011]). Recall from above that introducing extended fiber links in wireless networks may have a detrimental or even devastating impact on the throughput performance of distributed MAC protocols. Many open issues related to the design of low-cost physical layer components and layer-2/3 protocols and algorithms must be solved in order to render FiWi access networks commercially viable. In the following, we describe some of the major challenges and open issues of layer-2/3 protocols and algorithms in future RoF and R&F based FiWi access network architectures.

- **Integrated channel assignment and bandwidth allocation**
  To improve bandwidth-efficiency, powerful load balancing and reconfiguration techniques may be used in FiWi networks. In a FiWi access network, both bandwidth (of the fiber medium) and spectrum (of the wireless medium) must be dynamically allocated to provide better service to hotspot base stations and access points. The design of integrated dynamic bandwidth and spectrum allocation algorithms with good scalability is an open issue.
- **Integrated path selection**
  An open issue is the design of logical topologies of reconfigurable optical backhaul networks such that the number of required handovers especially for high-speed mobile customers can be decreased or avoided completely. In this context, the design of dynamic integrated path selection algorithms is an open issue.
- **Optical burst assembly and wireless frame aggregation**
  High-throughput next-generation 802.11n WLANs use two frame aggregation schemes known as aggregate MAC protocol data unit (A-MPDU) and aggregate MAC service data unit (A-MSDU), which can be used separately or jointly to increase the MAC throughput. EPON does not support frame aggregation. The benefit of one-level and/or two-level aggregation in EPON and converged EPON-WLAN networks is an open issue. To provide quality-of-service (QoS) and bandwidth-efficiency of FiWi networks, the design of hierarchical optical burst assembly and wireless frame aggregation schemes represents a promising avenue for future FiWi research.
- **Flow and congestion control**
  The bandwidth disparity between optical and wireless transmission media poses challenges at the network level. Traffic transiting from the optical to the wireless network could exceed its capacity and congest the wireless network, leading to significant buffer overflows and packet retransmissions. Developing appropriate flow and congestion control methods at the optical–wireless interface node to maximize network throughput is an open issue.

10.4 Summary

Hybrid optical–wireless FiWi networks form a powerful future-proof platform that provides a number of advantages. Introducing optical fiber into broadband wireless access networks helps relieve emerging bandwidth bottlenecks in today's wireless backhaul due to increasing traffic loads generated by new applications, e.g., iPhone. By simultaneously providing wired and wireless services over the same infrastructure, FiWi networks are able to consolidate (optical) wired and wireless access networks that are usually run independently of each other, thus potentially leading to major cost savings.

The few testbeds in existence or under development present a fraction of the possibilities existing in the integrated design of FiWi networks. By seamlessly converging optical and wireless access technologies, FiWi access networks hold great promise to support a plethora of future and emerging broadband services and applications on the same infrastructure. We briefly summarized the state of the art of enabling RoF and R&F technologies. Many open issues related to the design of low-cost physical layer components and layer-2 protocols and algorithms must be solved in order to render FiWi access solutions commercially viable. We highlighted two recent RoF and R&F testbeds and outlined future challenges and opportunities related to hybrid MAC protocols, integrated path selection, integrated channel assignment and bandwidth allocation, optical burst assembly and wireless frame aggregation, as well as flow and congestion control.

More interestingly, and certainly somewhat controversially, by paving all the way to and penetrating into homes and offices with high-capacity fiber and connecting wireless laptops and handhelds with high-throughput WiFi technologies to high-speed optical wired networks, FiWi networks give access to the ever increasing processing and storage capabilities of memory and CPUs of widely used desktops, laptops, and other wireless handhelds, e.g., Wii. Note that nowadays desktop and laptop computers commonly operate at a clock rate of 1 GHz with a 32-bit wide backplane, resulting in an internal flow of 2–8 Gb/s with today's limited hard drive I/O, while future desktops and laptops are expected to reach 100 Gb/s by 2010 (Green [2006]). At present, these storage and processing capabilities are quite often utilized only in part. After bridging the notorious first/last mile bandwidth bottleneck, research focus might shift from bandwidth provisioning to the exploitation of distributed storage and processing capabilities available in widely used desktops and laptops, especially as we are about to enter the Petabyte age with sensors everywhere collecting massive amounts of data (wir [2008]). An early example of this shift is the design of P2P on-line game architectures that have begun to increasingly receive attention, where players' computing resources are utilized to improve the latency and scalability of networked on-line games, whose groundbreaking technologies might also be used to realize the future 3D Internet. On the other hand, in-house computer facilities might be replaced with computer utilities as in-house generators were replaced with electrical utilities (Carr [2008]). Indeed, utility-supplied computing, e.g., Google, will continue to have an increasing impact on society and replace personal computer facilities unless new services and applications are developed that capitalize on them. Toward this end, it is important that FiWi networks are built using low-cost, simple, open, and ubiquitous technologies that allow all end-users to have broadband access and to create unforeseen services and applications that help stimulate innovation, generate revenue, and improve the quality of our every-day lives, while at the same time minimizing the associated technical, economical, societal, and personal risks.

11 Architectures

In this chapter, we review recent radio-over-fiber (RoF) and radio-and-fiber (R&F) based fiber-wireless (FiWi) network design proposals and discuss previously addressed challenges. Beside cell-based RoF networks, we briefly describe a number of FiWi network architectures, which can be classified based on their wireless access technologies: WiMAX or WiFi. Table 11.1 summarizes previously proposed WiMAX and WiFi based FiWi network architectures. While passive optical networks (PONs) can be widely found in FiWi networks, wireless mesh networks (WMNs) have been used rarely so far. As we will see shortly, different challenges have been addressed such as routing and wireless channel assignment, which can be performed completely either in the wireless domain by the base station (BS) or access point (AP), or by an optical network element, e.g., central office (CO) or optical line terminal (OLT). The level of provided quality-of-service (QoS) largely depends on the performance of the implemented routing and resource management algorithms, including bandwidth allocation and channel assignment algorithms with absolute or relative QoS assurances. Reconfiguration is another previously addressed challenging issue that involves resource management in the wireless and/or optical part. For instance, as explained in greater detail below, in the unidirectional ring/PON architecture of Table 11.1, highly loaded optical network unit-wireless gateways (ONU-WGs) may be assigned to a lightly loaded PON by tuning their optical transceivers to the wavelength assigned to the lightly loaded PON, resulting in a decrease of network congestion and packet latency.

11.1 Cellular architectures

Cellular networks used for fast-moving users, e.g., train passengers, suffer from frequent handovers when hopping from one BS to another one. The frequent handovers may cause numerous packet losses, resulting in a significantly decreased network throughput. An interesting approach to solving this problem is the use of an RoF network installed along the rail tracks in combination with the so-called *moving cell* concept (Lannoo *et al.* [2007]).

Table 11.1. Comparison of different FiWi network architectures.

Wireless access technology	Architecture	PON is used	WMN is used	Routing performed by	Channel assignment by	Level of QoS	Reconfiguration is provided	Protection is provided
WiMAX	Independent	Yes	No	BS	BS	Low	No	No
	Hybrid	Yes	No	ONU-BS	ONU-BS	Medium	No	No
	Unified connection-oriented	Yes	No	ONU-BS	ONU-BS	High	No	No
	Microwave-over-fiber	Yes	No	macro-BS	macro-BS	High	Yes	No
WiFi	Unidirectional ring	No	No	AP	AP	Medium	Yes	No
	Bidirectional ring	No	No	CO	AP	Low	No	Yes
	Hybrid star-ring	No	No	CO	AP	Medium	Yes	Yes
	Unidirectional ring/PON	Yes	Yes	ONU-WG	ONU-WG	High	Yes	Yes

BS: Base Station, AP: Access Point, WG: Wireless Gateway
ONU: Optical Network Unit, OLT: Optical Line Terminal, CO: Central Office

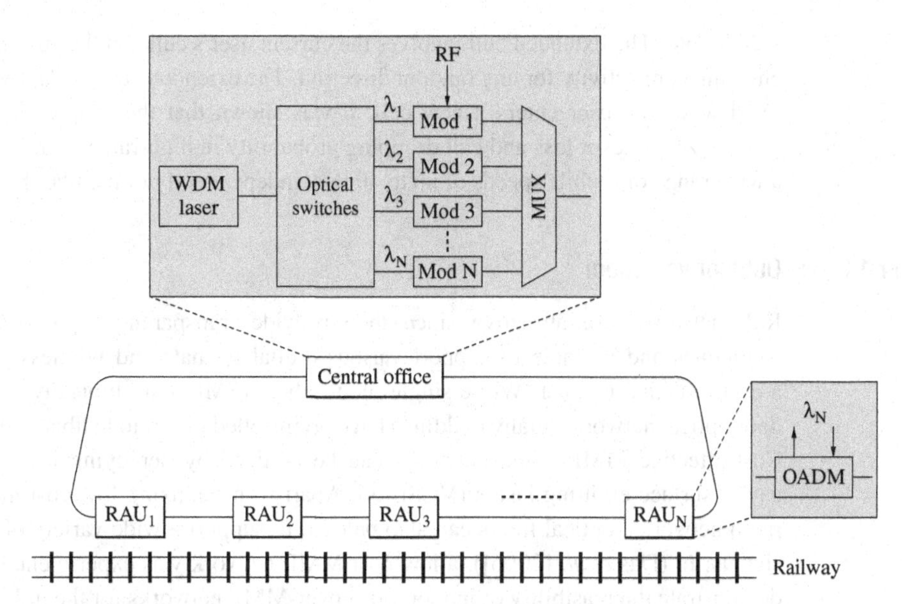

Figure 11.1 Moving cell-based RoF network architecture for train passengers. After Lannoo *et al.* (2007). ©2007 IEEE.

11.1.1 Moving cell

Figure 11.1 depicts the moving cell-based RoF network architecture for train passengers. An optical wavelength division multiplexing (WDM) ring interconnects the remote antenna units (RAUs) with the CO where all processing is performed. Each RAU deploys an optical add-drop multiplexer (OADM) fixed tuned to a separate wavelength channel. At the CO, a WDM laser generates the desired wavelengths, which are optically switched and passed to an array of radio frequency (RF) modulators, one for each RAU. The modulated wavelengths are multiplexed onto the optical fiber ring and received by each addressed RAU on its assigned wavelength. An RAU retrieves the RF signal and transmits it to the antennas of a passing train. In the upstream direction, the RAU receives all RF signals and sends them to the CO for processing. By processing the received RF signals, the CO is able to keep track of the train location and identify the RAU closest to the moving train. It then assigns downstream RF signals to the corresponding RAU such that the train and moving cells move along in a synchronous fashion.

11.1.2 Moving extended cell

Recently, the moving extended cell concept was proposed to provide connectivity for any possible direction (Pleros *et al.* [2009]). The fiber optic network becomes a means for speedy handoff between base stations that serve the mobile users. A hybrid frequency division multiplexing (FDM)/WDM network architecture was used to support the delivery of multiple RF channels in the 60 GHz frequency band over the same wavelength. The extended cell involves the current user's cell and the surrounding cells ensuring connectivity for any random direction. The extended cell is adaptively restructured when the user enters a new cell. It was shown that the proposed concept can provide zero packet loss and call dropping probability in high-rate wireless services for a wide range of mobile speeds of up to 40 m/s, independently of the fiber link distances.

11.1.3 Outdoor vs. indoor

RoF networks are attractive since they provide transparency against modulation techniques and are able to support various digital formats and wireless standards in a cost-effective manner. While single-mode fibers (SMFs) are typically found in outdoor optical networks, many buildings have preinstalled multimode fiber (MMF) cables. Cost-effective MMF-based networks can be realized by deploying low-cost vertical cavity surface emitting lasers (VCSELs). Apart from realizing low-cost microcellular radio networks, optical fibers can also be used to support a wide variety of other radio signals. In (Das *et al.* [2006]), a low-cost MMF network was experimentally tested to demonstrate the feasibility of indoor radio-over-MMF networks for the in-building coverage of second-generation and third-generation cellular radio networks as well as IEEE 802.11b/g wireless local area network (WLAN).

11.2 WiMAX-based architectures

11.2.1 Integrated EPON-WiMAX

The integration of Ethernet PON (EPON) and WiMAX access networks can be achieved in several ways. According to (Shen *et al.* [2007]), the following four architectures can be used:

- **Independent architecture**
  In this approach, WiMAX base stations serving mobile client nodes are attached to an ONU just like any other wired subscriber node. WiMAX and EPON networks are connected via a common standardized interface, e.g., Ethernet, and operate independently of each other.

- **Hybrid architecture**
  This approach introduces an ONU-base station (ONU-BS) that integrates the EPON ONU and WiMAX BS in both hardware and software. The integrated ONU-BS controls the dynamic bandwidth allocation of both ONU and BS.

- **Unified connection-oriented architecture**
  Similar to the hybrid architecture, this approach deploys an integrated ONU-BS. But instead of carrying Ethernet frames, WiMAX medium access control (MAC) protocol data units (PDUs) containing multiple encapsulated Ethernet frames are used. By carrying WiMAX MAC PDUs, the unified architecture can be run like

a WiMAX network with the ability to grant bandwidth finely using WiMAX's connection-oriented rather than EPON's queue-oriented bandwidth allocation.

- **Microwave-over-fiber architecture**
  In this approach, the WiMAX signal is modulated on a wireless carrier frequency and is then multiplexed and modulated together with the baseband EPON signal onto a common optical frequency (wavelength) at the ONU-BS. The central node consists of a conventional EPON OLT and a central WiMAX BS, called macro-BS. The OLT processes the baseband EPON signal, while the macro-BS processes data packets originating from multiple WiMAX BS units.

11.2.2 SuperMAN

Figure 11.2 depicts the SuperMAN architecture, which builds on an all-optical Ethernet-based access-metro network, described at length in (Maier *et al.* [2007]), extended with optical–wireless interfaces to next-generation WiFi and WiMAX networks. SuperMAN is a QoS-aware R&F access-metro network whose optical part consists of an IEEE 802.17 resilient packet ring (RPR) metro network that interconnects multiple WDM upgraded EPON access networks. Each WDM EPON has a tree topology, with the OLT at the tree root being collocated with one of the RPR ring nodes. The optical access-metro network lets low-cost PON technologies follow low-cost Ethernet technologies from access networks into metro networks by interconnecting the P OLTs with a passive optical WDM star subnetwork whose hub consists of a passive athermal wavelength-routing $P \times P$ arrayed-waveguide grating (AWG) in parallel with a wavelength-broadcasting $P \times P$ passive star coupler (PSC). The two types of optical–wireless interface in SuperMAN involve the integration of (i) RPR with WiMAX and (ii) EPON with next-generation WiFi (to be described in Chapter 16).

Figure 11.3 depicts the optical–wireless interface between an RPR metro ring node and a WiMAX access network, where the so-called integrated rate controller (IRC) (shaded in Fig. 11.3) plays a key role in integrating the two networks (Ghazisaidi *et al.* [2009]). It comprises a BS controller, a traffic class mapping unit, a central processing unit (CPU), and a traffic shaper. The IRC is used to seamlessly integrate both networks and jointly optimize the RPR scheduler and WiMAX downlink (DL)/uplink (UL) schedulers. The BS controller is responsible for handling incoming and outgoing WiMAX traffic, besides providing handover for subscriber stations (SSs) between different ring nodes. The traffic class mapping unit is able to translate the different WiMAX scheduling services and RPR traffic classes bidirectionally. The traffic shaper monitors the control rates of RPR traffic and performs traffic shaping according to RPR's fairness policies. The CPU synchronizes the aforementioned units and controls the RPR and WiMAX schedulers. More specifically, the CPU estimates the load of incoming traffic from different domains and synchronizes the schedulers based on traffic monitoring. RPR specifies the five traffic classes A0, A1, B-committed information rate (B-CIR), B-excess information rate (B-EIR), and C, while WiMAX specifies the following five scheduling services: unsolicited grant service (UGS), extended-real-time polling service

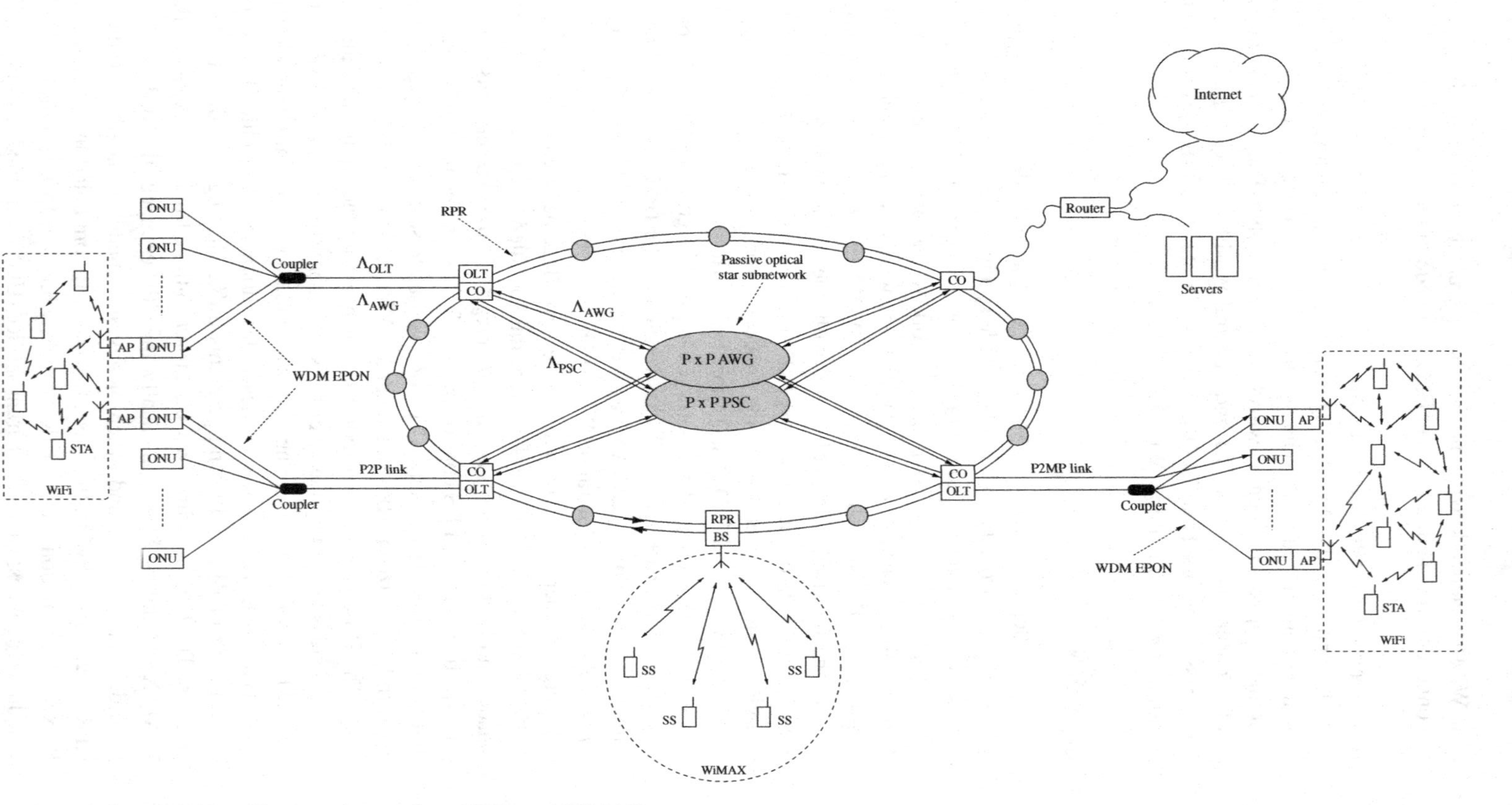

Figure 11.2 SuperMAN architecture: integration of RPR and WiMAX.

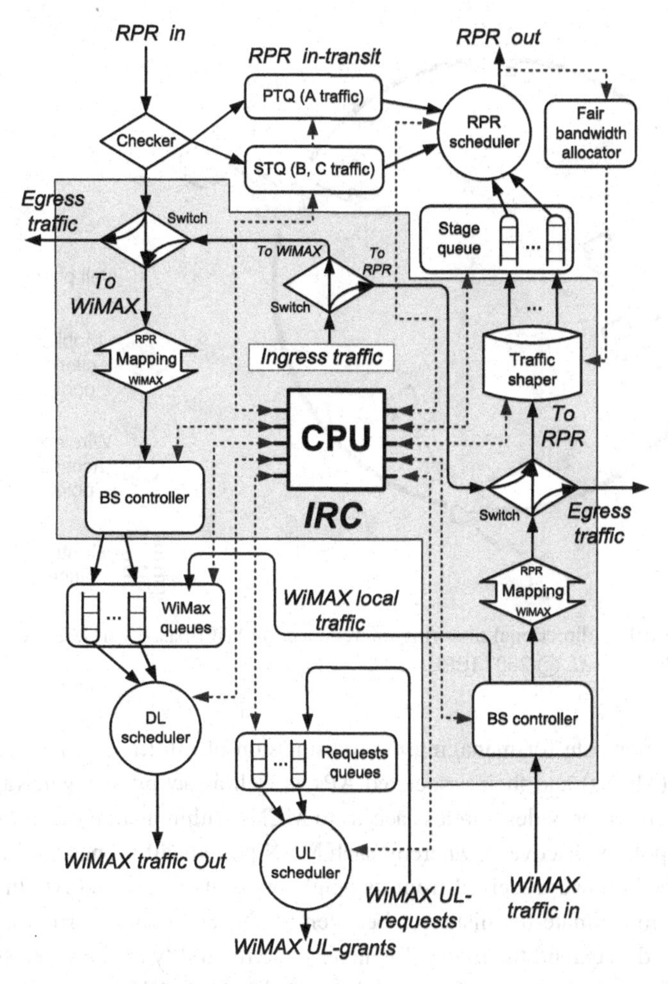

Figure 11.3 Optical–wireless interface between RPR and WiMAX networks.

(ErtPS), real-time polling service (rtPS), non-real-time polling service (nrtPS), and best effort (BE). RPR traffic classes and WiMAX scheduling services are mapped to each other in the order they are listed above.

11.3 WiFi-based architectures

Besides the aforementioned integration approaches of EPON and WiMAX networks, several other FiWi architectures based on WiFi technology have been studied, as described in the following.

11.3.1 Unidirectional ring

The network shown in Fig. 11.4 interconnects the CO with multiple WiFi-based wireless APs by means of an optical unidirectional fiber ring (Muralidharan *et al.* [2007]).

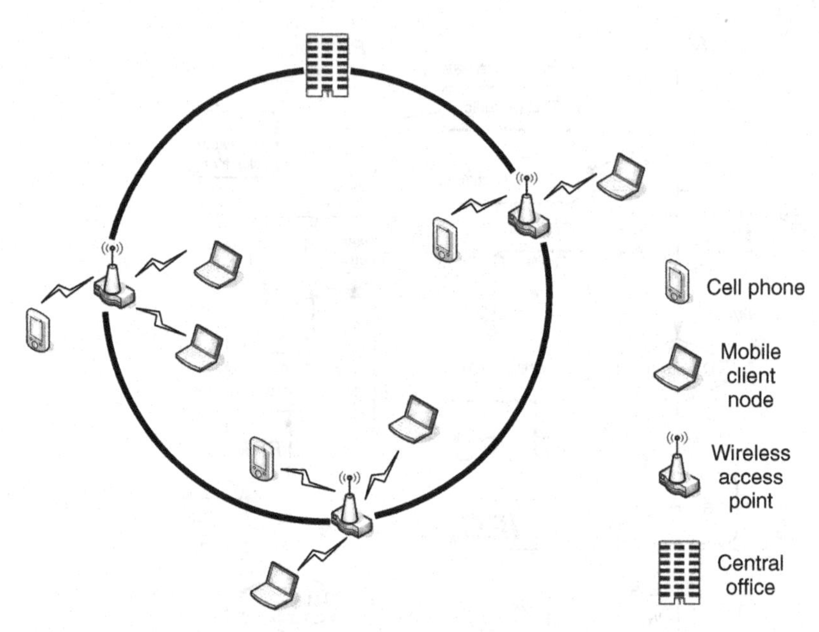

Figure 11.4 Optical unidirectional fiber ring interconnecting WiFi-based wireless access points. After Muralidharan *et al.* © 2007 IEEE.

The CO is responsible for managing the transmission of information between mobile client nodes (MCNs) and their associated APs as well as acting as a gateway to other networks. Each AP provides wireless access to MCNs within its range. All MCNs take part in the topology discovery, whereby each MCN periodically sends the information about the beacon power received from its neighbors to its associated AP. In doing so, APs are able to estimate the distances between MCNs and compute routes. Multihop relaying is used to extend the range. To enhance the reliability of the wireless link, the CO sends information to two different APs (path diversity). The proposed implementation can support advanced path diversity techniques that use a combination of transmission via several APs and multihop relaying, e.g., cooperative diversity or multihop diversity. Consequently, the CO must be able to assign channels quickly and efficiently by using one or more wavelength channels on the fiber ring to accommodate multiple services such as WLAN and cellular radio network.

11.3.2 Bidirectional ring

Figure 11.5 shows a two-level bidirectional path-protected ring (BPR) architecture for dense WDM (DWDM)/subcarrier multiplexing (SCM) broadband FiWi networks (Lin *et al.* [2003]). In this architecture, the CO interconnects remote nodes (RN) via a dual-fiber ring. Each RN cascades APs through concentration nodes (CNs), where each AP offers services to MCNs. For protection, the CO is equipped with two sets of devices (normal and standby). Each RN consists of a protection unit and a bidirectional wavelength add-drop multiplexer based on a multilayer dielectric interference filter.

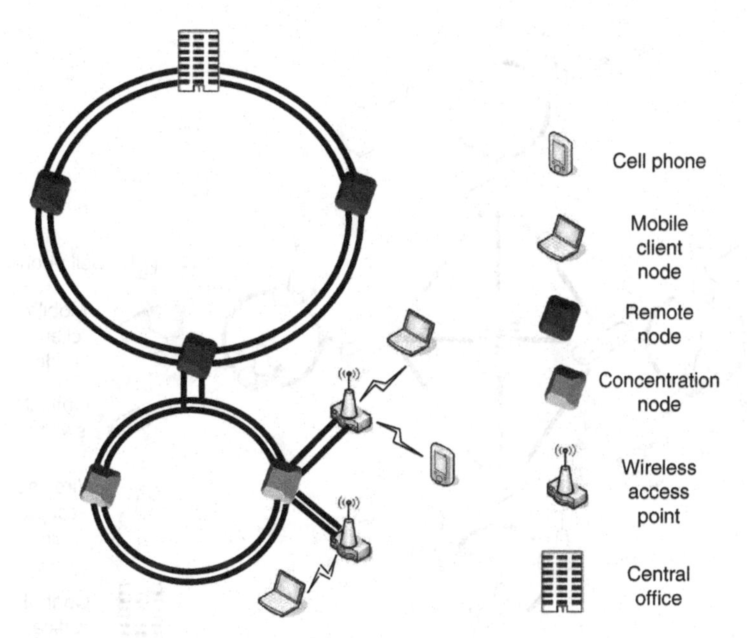

Figure 11.5 Optical interconnected bidirectional fiber rings integrated with WiFi-based wireless access points. After Lin *et al.* (2003). ©2003 IEEE.

Each CN contains a protection unit. The AP comprises an optical transceiver, a protection unit, up/down RF converters, and a sleeve antenna. Each AP provides channel bandwidth of at least 5 MHz and covers up to 16 MCNs by means of FDM. Under normal operating conditions, the CO transmits downstream signals in the counter-clockwise direction via RNs and CNs to the APs. If a fiber cut occurs between two RNs or between two CNs, their associated controllers detect the failure by monitoring the received optical signal and then switch to the clockwise protection ring. If a failure happens at an AP, the retransmitted signals are protection switched through other optical paths by throwing an optical switch inside the affected AP. This architecture provides high reliability, flexibility, capacity, and self-healing properties.

11.3.3 Hybrid star-ring

Figure 11.6 depicts a hybrid FiWi architecture that combines optical star and ring networks (Bhandari and Park [2006]). Each fiber ring accommodates several WiFi-based APs and is connected to the CO and two neighboring fiber rings via optical switches. The optical switches have full wavelength conversion capability and interconnect the APs and CO by means of shared point-to-point lightpaths. The network is periodically monitored during prespecified intervals. At the end of each interval, the lightpaths may be dynamically reconfigured in response to varying traffic demands. When traffic increases and the utilization of the established lightpaths is low, the load on the existing lightpaths is increased by means of load balancing. Otherwise, if the established lightpaths are heavily loaded, new lightpaths need to be set up, provided that enough

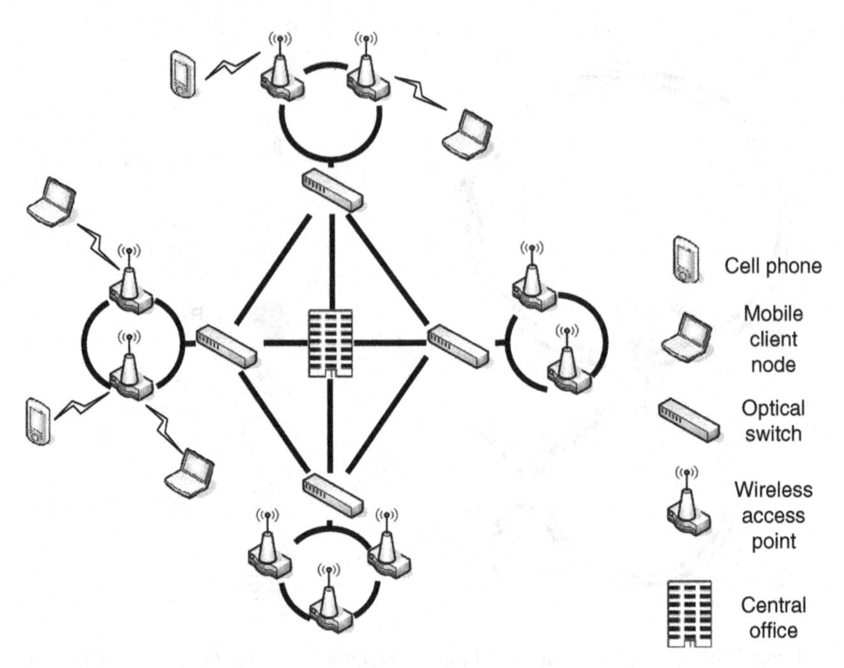

Figure 11.6 Optical hybrid star-ring network integrated with WiFi-based wireless access points. After Bhandari and Park (2006). ©2006 IEEE.

capacity is available on the fiber links. In the event of one or more link failures, the affected lightpaths are dynamically reconfigured using the redundant fiber paths of the architecture.

11.3.4 Unidirectional ring-PON

The FiWi network proposed in (Shaw *et al.* [2007a]) consists of an optical WDM backhaul ring with multiple single-channel or multichannel PONs attached to it, as shown in Fig. 11.7. More precisely, an OADM is used to connect the OLT of each PON to the WDM ring. Wireless gateways are used to bridge PONs and WMN. In the downstream direction, data packets are routed from the CO to the wireless gateways through the optical backhaul and are then forwarded to the MCNs by wireless mesh routers. In the upstream direction, wireless mesh routers forward data packets to one of the wireless gateways, where they are then transmitted to the CO on one of the wavelength channels of the optical backhaul WDM ring, as each PON operates on a separate dynamically allocated wavelength channel. Since the optical backhaul and WMN use different technologies, an interface is defined between each ONU and the corresponding wireless gateway in order to monitor the WMN and perform route computation taking the state of wireless links and average traffic rates into account. When the traffic demands surpass the available PON capacity, some of the single-channel time division multiplexing (TDM) PONs may be upgraded to WDM PONs. If some PONs are heavily loaded and others have less traffic, some heavy-loaded ONUs may be assigned to a lightly

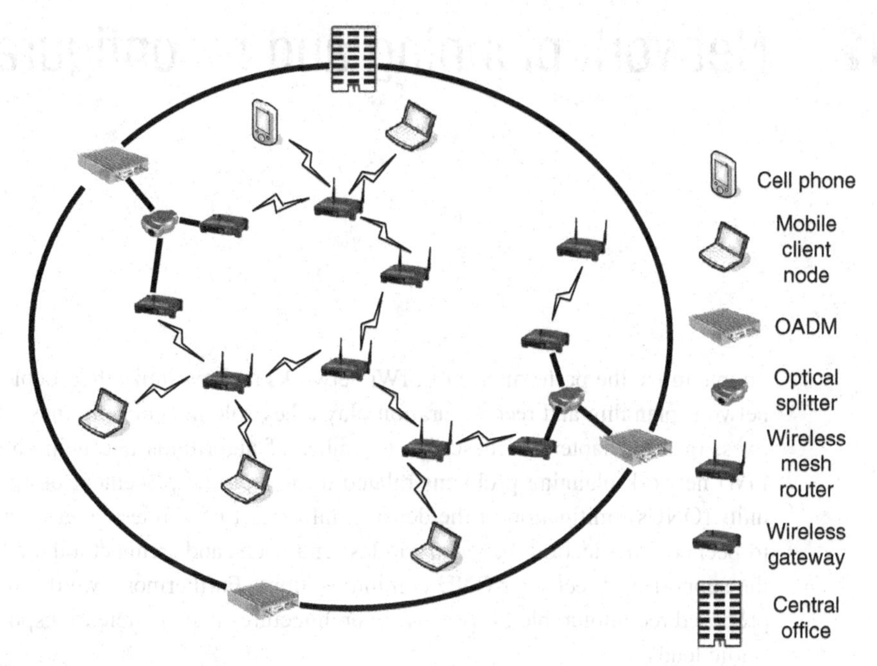

Figure 11.7 Optical unidirectional WDM ring interconnecting multiple PONs integrated with a WiFi-based wireless mesh network. After Shaw *et al.* (2007a). ©2007 IEEE.

loaded PON by tuning their optical transceivers to the wavelength assigned to the lightly loaded PON. This architecture provides cost-effectiveness, bandwidth efficiency, wide coverage, high flexibility, and scalability. In addition, the reconfigurable TDM/WDM optical backhaul helps reduce network congestion and average packet latency by means of load balancing. Moreover, the dynamic allocation of radio resources enables cost-effective and simple handovers.

11.4 Summary

We briefly summarized the state of the art of FiWi network architectures and previously addressed challenges. The few testbeds in existence or under development present a fraction of the possibilities existing in the integrated design of FiWi broadband access networks. FiWi networks become rapidly mature and give rise to new powerful access network solutions and paradigms. We have observed that research and development of future FiWi network architectures and protocols have made significant progress, but many open issues mostly related to the design of low-cost components, integrated routing, end-to-end service differentiation, and resiliency must be solved in order to render FiWi access solutions commercially viable.

12 Network planning and reconfiguration

To maximize the performance of FiWi networks and minimize their deployment costs, network planning and reconfiguration play a key role in achieving these design objectives. In this chapter, we describe a number of algorithms that help solve important FiWi network planning problems related to the optimal placement of optical network units (ONUs), mitigation of the detrimental impact of wireless interferences for peer-to-peer communications between wireless end-users, and architectural modifications for the support of direct inter-ONU communications. Furthermore, we discuss previously proposed reconfigurable FiWi network architectures that are able to respond to varying traffic loads.

12.1 ONU placement

The optimal placement of ONUs is an important design objective of FiWi networks due to the fact that the cost of laying optical fiber is significantly higher than that of devices attached to either end of the optical fiber, e.g., optical line terminal (OLT).

Several heuristics to solve the problem of optimally placing ONUs in a FiWi access network consisting of a passive optical network (PON) in tandem with a WiFi- or WiMAX-based wireless mesh network (WMN) were studied in (Sarkar *et al.* [2008]). The first proposed heuristic is a greedy algorithm that aims at finding a suitable placement of multiple ONUs to minimize the average Euclidean distance between wireless end-users and their closest ONU, i.e., this heuristic targets only the wireless front-end and does not take the fiber layout of the optical backhaul into account. The greedy algorithm starts with a given distribution of wireless end-users, which might be randomly or deterministically chosen, and consists of two phases. In the first phase, the algorithm identifies the primary ONU (i.e., closest ONU) for each wireless end-user. In the second phase of the algorithm, for each primary ONU a set of wireless end-users is obtained such that the distances between the ONU and its wireless end-users are minimized. The performance of the proposed greedy algorithm was evaluated in the Wildhorse neighborhood of North Davis, California, under the assumption of random and deterministic ONU placement. The presented results show that the greedy algorithm performs better for deterministic than for random ONU placement. The second proposed heuristic optimizes the placement of ONUs by means of *simulated annealing*, a widely used combinatorial optimization technique. For a given initial ONU distribution, this heuristic perturbs one of the high-cost ONUs to achieve a cost improvement in terms of average Euclidean distance between ONUs and wireless end-users. For each successful perturbation, wireless end-users are re-assigned to their nearest ONU. The reported results show that simulated annealing is able to achieve an overall cost improvement of typically close to 5% compared with the aforementioned greedy algorithm. Unlike the two previous algorithms, the third proposed heuristic also considers the cost of the fiber layout from ONUs to the OLT as well as interference in the wireless front-end. Toward this end, the proposed heuristic deploys the wireless access points (or base stations) uniformly, determines their transmission radius, and assigns channels to them such that the carrier-to-interference ratio does not drop below a certain threshold. After deploying the access points, the heuristic determines the number of ONUs needed to support the access points and deploys the ONUs according to the above described greedy algorithm. Finally, the heuristic finds a minimum-cost spanning tree to lay fiber from the OLT to all ONUs.

A powerful mixed integer programming (MIP) model for the optimum placement of ONUs and WiMAX base stations (BSs) in a PON/WiMAX-based FiWi network was developed in (Sarkar *et al.* [2007b, 2009]). The presented MIP model minimizes the number of deployed ONUs and BSs by saving costs and reducing the interference among multiple BSs and nearby wireless users. Specifically, the objective is to minimize the total installation cost of all ONUs and BSs required, including the cost of connecting the BSs to an ONU by means of optical fiber. Furthermore, the MIP problem takes into account a wide range of important constraints with regard to user assignment, ONU installation, BS installation, capacity, channel assignment, as well as signal quality and interference. More precisely, each wireless user is associated with at most one BS and the distance between a given wireless user and its associated BS must be within the transmission range of that BS. Moreover, the number of channels assigned to each BS is large enough to serve its wireless users, while not exceeding the upper bound of channels assigned to any BS. Each BS must be connected to only one ONU and the capacity of each ONU is large enough to serve all traffic generated by its attached BS. Finally, the quality of the wireless signal has to be at least the threshold of an acceptable carrier-to-interference ratio. For tractability, the authors used *Lagrangean relaxation* to relax some of the constraints and solve the MIP problem with acceptable accuracy. The presented simulation results indicate that the ONU/BS placement problem is quite sensitive to a number of parameters, e.g., number of available wireless channels and wireless user coverage ratio.

12.2 Inter-ONU communications

12.2.1 Peer-to-peer communications

In FiWi networks, traffic may go from wireless end-users to the Internet or from one wireless client to another wireless client. In the former case, the throughput of FiWi networks is limited by the bandwidth bottleneck and interferences of communications

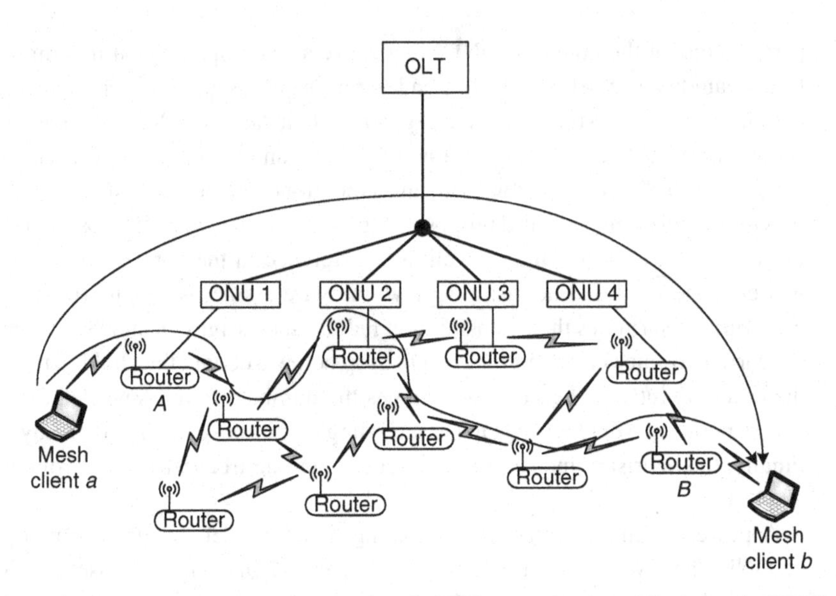

Figure 12.1 Peer-to-peer communications in FiWi networks. After Zheng *et al.* (2009b). ©2009 IEEE.

in the wireless subnetwork. In the latter case, the bimodal nature of FiWi networks provides an opportunity to mitigate the detrimental impact of wireless interferences on the network throughput for peer-to-peer communications between two wireless end-users. In FiWi networks, peer-to-peer traffic can be carried in two ways: (*i*) either through the wireless front-end only, or (*ii*) along a wireless–optical–wireless path (Zheng *et al.* [2009b]). As shown in Fig. 12.1, suppose that wireless mesh client *a* wants to communicate with mesh client *b*. The resultant peer-to-peer traffic can be routed either through the multi–hop wireless path within the wireless subnetwork or through the wireless–optical–wireless path, whereby traffic is first sent from mesh client *a* via its associated mesh router *A* to its closest ONU 1 and is then forwarded upstream to the OLT. Subsequently, the OLT broadcasts the traffic downstream to all ONUs. Each receiving ONU determines to discard or forward the traffic according to the location of the destination mesh client. In our example, ONU 4 is closest to destination mesh client *b* and thus ONU 4 will forward the arriving traffic to *b* via its associated mesh router *B*, while all remaining ONUs discard the traffic coming from the OLT. Note that traffic forwarded along the wireless–optical–wireless path alleviate interferences in the wireless front-end, thereby allowing more traffic to be carried in the wireless segment and resulting in an increased network throughput. Also note that the capacity of the PON is much higher than that of the wireless front-end such that peer-to-peer traffic can be easily carried in the optical backhaul without suffering from any serious throughput penalty.

Clearly, to exploit the benefits of steering peer-to-peer traffic along optical–wireless–optical paths, the placement of ONUs has a great impact on the achievable throughput of FiWi networks. In (Zheng *et al.* [2009b]), the ONU placement was optimized with the objective to maximize the throughput of FiWi networks for peer-to-peer communications, assuming that single-radio single-channel wireless routers are deployed and

ONUs can communicate with each other by sending traffic to the OLT, which in turn broadcasts traffic back to all ONUs. Given the location of the wireless mesh routers, the optimal placement of K ONUs was found with the objective of minimizing the total number of required wireless hops. Once the location of the K ONUs is fixed, each wireless node calculates the shortest path (in terms of hops) within the wireless mesh front-end to each ONU and selects the ONU with the minimal hop count. To solve this optimization problem, the authors applied a *tabu search* based heuristic, a widely used metaheuristic algorithm for solving combinatorial problems. The presented simulation results show that under different network sizes the proposed tabu search heuristic outperforms random and fixed ONU placement schemes in terms of achievable network throughput, especially for an increasing number of ONUs.

12.2.2 FiWi vs. WMN networks

To measure the network throughput gain in FiWi networks, a linear programming based routing algorithm was proposed in (Zheng *et al.* [2009a]). The objective of the routing algorithm is to route traffic in the FiWi network such that network throughput is maximized. The proposed optimal routing algorithm yields a bound on the throughput gain in FiWi networks. Extensive simulations were conducted to study the network throughput gain in FiWi networks under peer-to-peer traffic among wireless mesh clients and compare the achievable throughput gain with conventional WMNs without any optical backhaul. The presented simulation results show that the network throughput gain of FiWi networks is zero compared with traditional WMNs when all traffic is destined to the Internet, i.e., no peer-to-peer traffic, since the interference in the wireless front-end is the major bandwidth bottleneck. In other words, when all the traffic goes to the Internet, the throughput of FiWi networks is the same as that of traditional WMNs. However, with increasing peer-to-peer traffic the interferences in the wireless mesh front-end increase and the throughput of WMNs decreases dramatically. In contrast, the throughput of FiWi networks decreases to a much lesser extent for increasing peer-to-peer traffic due to the above described wireless–optical–wireless communications mode of FiWi networks. Thus, these findings show FiWi networks are particularly beneficial for supporting peer-to-peer communications among wireless mesh clients and are able to achieve a significantly higher network throughput than conventional WMNs.

12.2.3 Direct inter-ONU communications

The throughput-delay performance of FiWi networks for peer-to-peer communications can be further improved by means of direct inter-ONU communications. In conventional PONs, inter-ONU communications is done via the OLT. In (Li *et al.* [2010b]), a new wavelength division multiplexing/time division multiplexing (WDM/TDM) PON-based FiWi network architecture was proposed and investigated to enable direct inter-ONU communications without going through the OLT. The proposed WDM/TDM PON architecture requires major hardware extensions at the OLT, ONUs, and remote node. More specifically, the OLT is equipped with an array of fixed-tuned transmitters and an array of fixed-tuned receivers, each operating on a different dedicated wavelength channel, for transmitting and receiving data to and from the ONUs. Each ONU deploys a fixed-tuned transmitter and fixed-tuned receiver at dedicated upstream and downstream wavelength channels. In addition, each ONU uses a tunable transmitter and a fixed-tuned receiver for inter-ONU communications. The inter-ONU traffic from all ONUs are combined into one fiber by means of a passive combiner, which is attached to one of two input ports of an arrayed-waveguide grating (AWG). Both the combiner and AWG are located at the remote node of the WDM/TDM PON. The second input port of the AWG is used to interconnect the OLT with the ONUs for communications between the OLT and ONUs. The additional transceiver enables ONUs to send data directly to each other across the AWG rather than going through the OLT, resulting in an improved throughput-delay performance under peer-to-peer traffic. On the downside, it must be mentioned that the proposed architecture is quite complex and costly. Not only do the required tunable transmitters add to the complexity and cost of each ONU but also optical amplifiers might become necessary to compensate for the coupling and insertion losses of the inserted combiner and AWG, thereby violating the unpowered nature of PONs and cost-sensitivity of access networks in general.

12.3 Reconfiguration

In Section 11.3.4, we introduced an optical unidirectional WDM ring interconnecting multiple PONs integrated with a WiFi-based WMN (see Fig. 11.7). To adapt the considered FiWi network to traffic changes in different districts, reconfigurability was implemented in the optical backhaul (Shaw *et al.* [2007a, 2008]). Instead of overprovisioning based on the peak traffic demand, it is desirable to reallocate bandwidth among multiple PONs by utilizing tunable transceivers at the ONUs. For instance, let us assume that two PONs share the same optical tree subnetwork and one PON is highly loaded while the other PON is not, then by tuning the transmitting and receiving wavelengths of a highly loaded ONU associated with the former PON the ONU is enabled to join the latter lightly loaded PON in order to achieve load balancing between the two PONs. Toward this end, in the central hub of the FiWi network a network terminal continually monitors the buffer depth of each OLT for the downstream traffic. As a PON becomes heavily loaded, as indicated by the corresponding OLT buffer depth, the heavily loaded PON is instructed to deregister some ONUs and reregister them to the lightly loaded PON(s). Before an ONU is deregistered, its packets queued at the OLT will first be emptied. After an ONU is deregistered, the incoming traffic to that ONU is temporarily stored at the network terminal until the reregistration is achieved. To reduce the overhead of slow transceiver tuning times during reconfiguration and increase network throughput, the buffer size may be increased and a higher threshold for the reconfiguration trigger may be set, at the expense of longer queueing delays in the buffer.

Note that the above described reconfigurable FiWi network is referred to as the MARIN hybrid optical–wireless access network in (Shaw *et al.* [2007b]). However, the same authors also used the term MARIN to denote a reconfigurable all-optical access-metro backhaul network, as explained next.

12.3.1 MARIN

The so-called *metro access ring integrated network (MARIN)* optically integrates hybrid TDM/WDM PONs into interconnected metro access ring networks by using optical reconfigurable and parametric wavelength conversion (PWC) devices (Shaw *et al.* [2006]). Specifically, MARIN consists of multiple interconnected metro access dense WDM (DWDM) rings. Each metro access ring has its own central office (CO) that schedules the distribution/collection of traffic to/from attached PONs, coordinates with COs of other metro access rings to forward metropolitan area network (MAN) traffic, and interfaces with the metro core network at the higher level of the network hierarchy. The CO deploys tunable lasers whose configuration is done according to given traffic loads in the metro access rings and attached PONs. Each PON in a given metro access DWDM ring is addressed on a separate set of wavelength channels. Two different types of network nodes are used in the metro access DWDM ring: (*i*) the MARIN gateway, and (*ii*) the MARIN switch. A MARIN gateway drops the wavelengths carrying downstream traffic destined for the attached PON subscribers. At a MARIN switch, wavelengths that carry in-transit MAN traffic are all-optically routed using a reconfigurable wavelength-selective switch (WSS) and PWC. PWC allows for the any-to-any conversion of multiple wavelengths (wavelength set) at the same time.

In the resultant all-optical MARIN network, PON resources can be shared and leveraged by metro access ring networks. More specifically, a MARIN gateway can dynamically share light sources that were originally used to serve only the attached access network, resulting in a more efficient utilization of network resources and improved network performance (Wong *et al.* [2007]). Figure 12.2 depicts the architecture of a MARIN gateway, which is able to utilize idle or over-provisioned tunable transmitters (TTs) of the attached WDM PON through a passive AWG wavelength router. The MARIN gateway is able to add/drop metro traffic on wavelength channel λ_L deployed in legacy single-channel ring networks, e.g., SONET/SDH or resilient packet ring (RPR). In MARIN, the legacy wavelength channel λ_L may also be used as a control channel for wavelength reservation in support of quality-of-service (QoS) and optical burst switching (OBS). In addition, the MARIN gateway is able to add metro traffic on any wavelength channel λ_M of the DWDM ring by using one of the available TTs and drop metro traffic on a dedicated home wavelength channel λ_1, whereby each MARIN gateway is assigned a different home wavelength channel for reception of metro ring traffic. A frequency-cyclic AWG in conjunction with a number of TTs are used to improve network scalability and flexibility. The frequency-cyclic AWG is used as a wavelength router to dynamically share TTs and wavelengths between the metro access ring and the attached WDM PON.

12.3.2 GROW-Net

The so-called *grid reconfigurable optical–wireless network (GROW-Net)* is a scalable municipal FiWi network architecture that adapts the street layout in a city and makes use of available dark fibers readily available in urban areas (Shaw *et al.* [2009], Wong *et al.* [2009]). The design objectives of GROW-Net are to provide scalability in

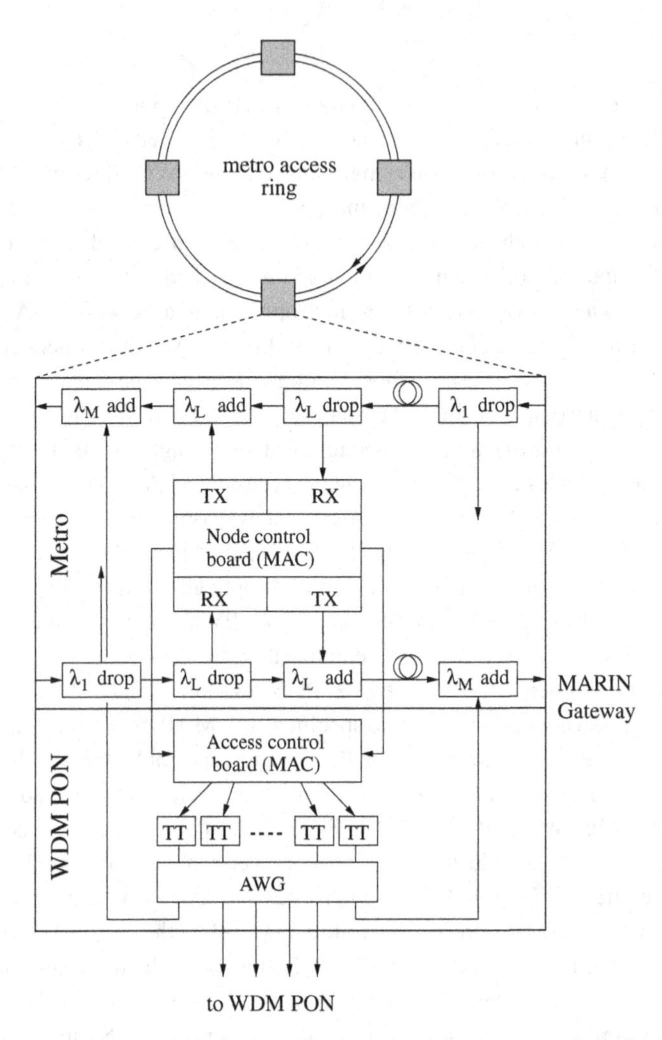

Figure 12.2 MARIN gateway architecture. After Wong *et al.* (2007). ©2007 IEEE.

terms of bandwidth and allow for infrastructure extensibility. The basic building block of GROW-Net is the grid-cell, which provides the backbone fiber connectivity to wireless gateways. The optical fiber backbone infrastructure is extensible using the proposed H-tree partitioning process, whereby at each stage of the process the spacing between two adjacent WMN routers is reduced by 50% and optical terminals are inserted and interconnected by means of optical WDM links that are laid out in an H-shape. In the downstream direction, the central hub of GROW-Net deploys a tunable laser, which can reach different optical terminals by tuning the laser to the respective wavelength channel supported by the intended optical terminal. Each optical terminal is equipped with a thin film filter (TFF) that lets pass only the corresponding wavelength channel of a given optical terminal. In the upstream direction, each optical terminal is made colorless by deploying a reflective semiconductor optical amplifier (RSOA) for remote modulation of a continuous wavelength signal sent by the central hub. GROW-Net is able to achieve increased fiber protection compared with conventional tree-based PONs due to its mesh topology.

12.4 Summary

We have reviewed several heuristics to solve the problem of optimally placing ONUs in FiWi networks that are based on integrated PON and WiFi- or WiMAX-based WMN networks. The presented heuristics aim at meeting a number of different objectives, including minimizing the average Euclidean distance between wireless end-users and their closest ONUs, determining the number of required ONUs, finding a minimum-cost spanning tree to lay fiber from the OLT to all ONUs, and reducing the interference among access points/base stations and nearby wireless users. We have seen that FiWi networks provide an opportunity to mitigate the detrimental impact of wireless interferences on the network throughput for peer-to-peer communications between wireless end-users without suffering from any serious throughput penalty on the optical backhaul. In fact, FiWi networks are particularly beneficial for supporting peer-to-peer communications among wireless mesh clients and are able to achieve a significantly higher network throughput than conventional WMNs. FiWi networks can effectively adapt to traffic changes in different areas by rendering the optical backhaul reconfigurable and sharing PON resources such as tunable transmitters in all-optically integrated metro access networks.

13 Techno-economic analysis

Recently, various fiber-wireless (FiWi) network architectures have been investigated by integrating different optical and wireless technologies. While introducing optical fiber at higher network layers, e.g., aggregation layer, helps alleviate emerging bandwidth bottlenecks, the last hop is expected to be wireless for ubiquity and convenience, e.g., low-cost wireless home area networks (HANs) (He *et al.* [2008]). Between these two FiWi network hierarchy levels lies the "sweet spot" where optical technologies interface with their wireless counterparts. Two important sweet-spot technologies that play a key role in emerging FiWi networks are Ethernet passive optical network (EPON) and WiMAX. Clearly, EPON and WiMAX networks may be cascaded, as proposed in (Shen *et al.* [2007]). However, given the similarities of EPON and WiMAX (e.g., point-to-multipoint topology with a central control station performing dynamic bandwidth allocation (DBA) by means of centralized polling and scheduling) we argue that the two technologies are more likely to target the same network segment rather than being cascaded to cover different network segments. In other words, we expect that network operators will make a choice between EPON and WiMAX depending on a number of factors, e.g., right-of-way, and elaborate on the techno-economic comparison of the two technologies.

For the comparison of wired and wireless network technologies various techno-economic evaluation techniques have been proposed, as we will see shortly. During the last decade, the techno-economic evaluation of various network technologies has been an active research area. To meet the different requirements of emerging network services, a service migration cost analysis was presented in (Shayani *et al.* [2008]). The cost modeling of the migration from best-effort access networks to multi-service Quality-of-Service (QoS) enabled access networks based on Ethernet and asynchronous transfer mode (ATM) was proposed in Monath and Kind [2007]. The obtained results show that deployment cost savings can be achieved by using Ethernet-based access network architectures. It is important to note that most of the previous techno-economic evaluations focused either on optical fiber only (e.g., Weldon and Zane [2003] and Tran *et al.* [2005]) or wireless only network architectures (e.g., Niyato and Hossain [2007]). Up to date, only a few preliminary techno-economic evaluations of FiWi networks have been reported. A cost comparison of VDSL and a FiWi architecture consisting of cascaded EPON and WiMAX networks was carried out in (Lin *et al.* [2007b]). The obtained results indicate the superior cost-efficiency of FiWi networks over conventional VDSL solutions. In (Chowdhury *et al.* [2008]), a deployment cost comparison of wired (i.e., xDSL and cable modem), optical fiber, WiFi, and integrated EPON and WiMAX/WiFi network architectures was made. The reported results show that a hybrid FiWi network architecture (consisting of EPON and WiMAX) represents a cost-effective solution for future broadband urban area networks. Different FiWi network design heuristics were investigated in terms of processing time, complexity, and installation cost in (Sarkar *et al.* [2007a]). The optimum real-estate cost deployment of optical network units (ONUs) in integrated FiWi networks was studied in (Sarkar *et al.* [2006]) and (Sarkar *et al.* [2007b]). Despite these preliminary studies, a more thorough techno-economic evaluation of FiWi networks is necessary in order to gain deeper insights into the design, configuration, and performance optimization of emerging FiWi networks that are based on EPON and/or WiMAX technologies. Toward this end, we are going to first introduce the various components of total cost of network ownership. We then perform a comparative techno-economic analysis of EPON and WiMAX and present insightful results, including recently standardized IEEE 802.3av 10 Gb/s EPON and 802.16m 1 Gb/s WiMAX networks. We refer the interested reader to (Ghazisaidi and Maier [2010]) for more detailed information.

13.1 Total cost of network ownership

The total network deploying expenditures for network operators are called *total cost of network ownership (TCO)*. Typically, TCO is categorized into: (i) capital expenditures (CAPEX) and (ii) operational expenditures (OPEX). In this section, we provide a general overview of the most important TCO components widely considered in previous techno-economic network studies.

13.1.1 CAPEX

CAPEX consist of initial network equipment and network installation costs, network infrastructure costs (e.g., cabling and right-of-way), and network management system. Additionally, CAPEX cover the upgrading and protection of spare network equipment and installation costs. The first-time installation cost is covered by CAPEX, in case it should be done by network operators. We note that the first-time installation is usually done by the equipment vendors. Non-telecom costs such as building and furniture costs are usually considered part of network CAPEX (Machuca [2006]).

13.1.2 OPEX

OPEX comprise network operation, administration, and maintenance (OAM) costs. More specifically, OPEX cover the network power consumption and equipment cooling, troubleshooting, repairing, service (i.e., service provisioning and management), and human resource costs, e.g., wages and salaries. The non-telecom costs, e.g., room air-conditioning and heating, are defined as network OPEX (Machuca [2006]). According to (Verbrugge *et al.* [2005]), the network OPEX may be classified as follows:

- **OPEX for setting up a network:** which include in-advance planning cost, e.g., initial network planning cost and travel cost for contacting and negotiating with different equipment vendors.
- **OPEX to operate an existing network:** which comprise the continuous cost of infrastructure (e.g., power consumption and cooling costs), maintenance cost, failure reparation cost, provisioning and service management cost, pricing and billing cost, operational network planning cost (e.g., day-to-day planning, re-optimization, and upgrade planning), and marketing cost.
- **Non-telecom OPEX:** which include the continuous infrastructure cost (e.g., leasing) and specific administration cost, such as employees' payments and human resource operations.

13.2 Comparative analysis of EPON and WiMAX

In this section, we introduce a novel techno-economic model, where the initial network deployment costs are considered as network CAPEX, while power consumption, cooling, and repair costs are considered as network OPEX. In our approach, we do not consider the optimum real-estate cost deployment of EPON ONUs, since it is expected that the ONUs are located according to given residential or business subscribers' preferences. Instead, three different terrain types (i.e., urban, suburban, and rural) are defined to investigate the impact of network population density. Various terrain types and population densities are considered to evaluate the performance of our proposed techno-economic model for both EPON and WiMAX. Additionally, we consider the high-speed IEEE standards 802.3av 10 Gb/s EPON and 802.16m, also known as ultra-high-bandwidth WiMAX, which aims at providing 1 Gb/s for fixed subscriber stations (SSs) and 100 Mb/s for mobile SSs.

13.2.1 Techno-economic model

Figure 13.1 illustrates our proposed techno-economic model for the comparative analysis of EPON and WiMAX. It consists of the following modules:

- **Scenario description:** this defines various network deployment scenarios and terrain type conditions (i.e., urban, suburban, and rural).
- **Technological constraints:** this module determines the technological limitations of a given scenario, such as the maximum distance between OLT and ONUs in EPON.
- **Initial network infrastructure:** this designs the initial network infrastructure of a given scenario with given constraints of the applied technology.
- **Cost-modeling techniques:** this module includes the cost-modeling methods used in the subsequent cost-efficient network design.
- **Cost-efficient network design:** this module modifies or redesigns the initial network infrastructure making use of the cost-modeling techniques module.
- **Cost calculation:** this calculates the network costs, which are categorized into: (i) CAPEX and (ii) OPEX.

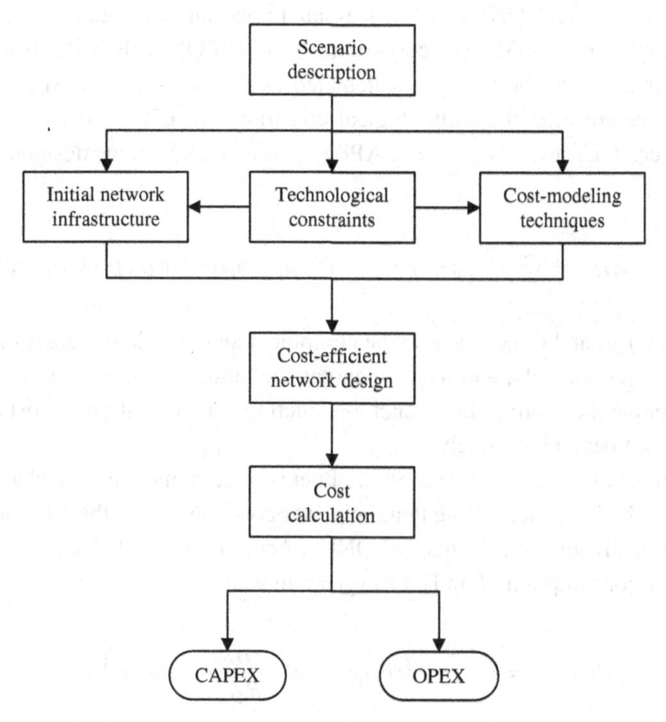

Figure 13.1 Techno-ecnomic model.

13.2.2 Techno-economic evaluation

In the following, the proposed techno-economic evaluation techniques for both EPON and WiMAX are discussed in greater detail.

EPON

In previous studies (Weldon and Zane [2003], Tran *et al.* [2005], Lin *et al.* [2007b]), ONUs were located equidistantly l meters from each other. In contrast, we allow ONUs to be uniformly randomly distributed in a square cell, whose dimensions are determined based on the given terrain type (urban, suburban, or rural), whereby $l_{rural} > l_{suburban} > l_{urban}$ denote the maximum distance of a pair of adjacent ONUs for the respective terrain type. Unlike previous studies, our ONU localization approach allows us to examine more realistic EPON deployment scenarios where ONUs may be placed at different distances, depending on the terrain type and population density. Note that our ONU placement scheme includes the equidistant scenario as a special case.

Different network configurations can be designed for the above-mentioned scenarios subject to the technological constraints of EPON. Next, we introduce the maximum cost-sharing approach as a cost-modeling technique for our proposed techno-economic model. The maximum cost-sharing technique aims at minimizing the length of the required distribution fibers (between remote node (RN) and ONUs). In this approach, the position of the OLT is fixed and the distance between the optical line terminal (OLT) and ONUs is used as an important parameter in cost modeling (Weldon and Zane [2003]). The costs of deploying EPON with ONUs being located at different ranges from the OLT are calculated as follows. In our cost-efficient network design, we consider the technological constraints of EPON, such as maximum distance between OLT and ONUs. The CAPEX (given in US $) of the designed EPON is given by

$$C_{EPON} = C_{OLT} + C_{OLT-RN} + C_{RN} + C_{RN-ONU} + n_{ONU} \cdot C_{ONU}, \qquad (13.1)$$

where C_{OLT}, C_{RN}, and C_{ONU} denote the equipment and installation costs of the OLT, RN, and ONU, respectively, and n_{ONU} denotes the number of ONUs. C_{OLT-RN} and C_{RN-ONU} denote the cabling (i.e., trenching, ducting, and optical fiber) costs of feeder and distribution fibers, respectively.

In our approach, the power consumption of network equipment is considered a major permanent OPEX component. Note that the power consumption of the OLT depends on the requested bandwidth of the attached ONUs. According to (Gladisch *et al.* [2008]), the total power consumption of an EPON (given in watt) is given by

$$P_{EPON} = n_{ONU} \cdot \left(P_{ONU} + 2 \cdot \frac{r_{ONU}}{R_{OLT}} \cdot P_{OLT} \right), \qquad (13.2)$$

where P_{ONU} and P_{OLT} denote the power consumption of an ONU and OLT, respectively. In Eq. (13.2), the capacity of EPON is denoted by R_{OLT}, while r_{ONU} denotes the requested average bandwidth of each ONU. We note that the factor 2 is used to account for the power consumed for both service provisioning and heat dissipation (Gladisch *et al.* [2008]).

In previous studies (Gladisch *et al.* [2008], Baliga *et al.* [2008]), only power consumption was considered in the OPEX calculation, however another important OPEX component of EPON networks is the repairing cost. Typically, two types of failures can take place in an EPON: (i) a node failure might occur at the OLT, RN, or ONU; and (ii) a fiber cut might occur in feeder and distribution fibers. The probability of an EPON network failure due to a single node failure or fiber cut is given by

$$Pr_{SF-E} = p_{NE} \times \left(\frac{1 + 1 + \binom{n_{ONU}}{1}}{\binom{n_{ONU}+2}{1}} \right) + p_{NE} \times \left(\frac{1 + \binom{n_{ONU}}{1}}{\binom{n_{ONU}+1}{1}} \right), \qquad (13.3)$$

where p_{NE} denotes the probability of a network element (i.e., node or fiber) failure, whereby $0 \leq p_{NE} \leq 1$. In Eq. (13.3), the first fraction calculates the probability that the OLT, RN, or one of the ONUs fails and the second fraction computes the probability of a feeder or distribution fiber cut. Note that the node and fiber failure events are mutually exclusive. The probability of an EPON network failure due to multiple node failures and/or fiber cuts is given by

$$
\begin{aligned}
Pr_{MF-E} = & p_{NE}^{f_n} \times \left(\frac{\binom{n_{ONU}}{f_n} + 2 \cdot \binom{n_{ONU}}{f_n - 1} + \binom{n_{ONU}}{f_n - 2}}{\binom{n_{ONU}+2}{f_n}} \right) \\
& + p_{NE}^{f_f} \times \left(\frac{\binom{n_{ONU}}{f_f} + \binom{n_{ONU}}{f_f - 1}}{\binom{n_{ONU}+1}{f_f}} \right),
\end{aligned}
\tag{13.4}
$$

where f_n and f_f denote the number of node failures and fiber cuts, respectively. In Eq. (13.4), the probability of node failures and fiber cuts is given in the first and second fractions, respectively, whereby $f_n \geq 2$ and $f_f \geq 1$.

The cost of repairing a single or multiple failures in an EPON is given by

$$C_{R-EPON} = \sum_{\forall i} [Pr_i \times (\alpha \cdot CAPEX(i))], \tag{13.5}$$

where i denotes the failed EPON node or fiber, Pr_i denotes the probability of failure for equipment i, deduced from Eqs. (13.3) and (13.4), $CAPEX(i)$ denotes the initial equipment and installation cost of device i, and α denotes the repairing or replacing factor, whereby $0 < \alpha < 1$ if the failed device can be repaired and $\alpha = 1$ if it has to be replaced.

Hence, the OPEX (given in US $) of the designed EPON network is given by

$$O_{EPON} = C(P_{EPON}) + C_{R-EPON}, \tag{13.6}$$

where $C(P_{EPON})$ denotes the cost of power consumed by the EPON.

WiMAX

In our WiMAX scenarios, we consider outdoor WiMAX antennas in order to provide a fair comparison between WiMAX and EPON networks. Three different terrain types (i.e., urban, suburban, and rural) are defined to provide realistic modeling conditions, whereby $PL_{urban} > PL_{suburban} > PL_{rural}$ denote the wireless path loss of the corresponding terrain type. According to (Ghazisaidi *et al.* [2009]), for all three terrain types the path loss PL between a given SS in range L of the BS is given by

$$PL = A + 10\gamma \log_{10}(L/L_0) + s + \Delta PL_f + \Delta PL_{h_{SS}}. \tag{13.7}$$

In Eq. (13.7), $A = 20 \log_{10}(4\pi L_0/\lambda)$, where λ and L_0 denote the wavelength (given in meter) and reference distance between the SS and BS with $L > L_0 = 100$ m, respectively. Further, γ denotes the path loss exponent which is given by

$$\gamma = a - b \cdot h_{BS} + c/h_{BS}, \tag{13.8}$$

where h_{BS} denotes the height of the BS antenna (10 m $< h_{BS} <$ 80 m) and the value of the parameters a, b, and c are set differently for each terrain type according to Table 13.1. In Eq. (13.7), s (given in dB) denotes the shadow fading whose value depends on the terrain type and is given by

$$s = y(\mu_\sigma + z\sigma_\sigma), \tag{13.9}$$

where y and z denote zero-mean Gaussian variables of unit standard deviation $N[0, 1]$ and the parameters μ_σ and σ_σ are set to different values according to Table 13.1

Table 13.1. Path loss parameter values for various terrain types.

Parameter	Urban	Suburban	Rural
a	4.6	4	3.6
b	0.0075	0.0065	0.005
c	12.6	17.1	20
μ_σ	10.6	9.6	8.2
σ_σ	2.3	3.0	1.6

depending on the terrain type. The two correction terms ΔPL_f and $\Delta PL_{h_{SS}}$ in Eq. (13.7) denote the frequency and antenna height of the SS, respectively. For all three terrain types, ΔPL_f is fixed to $\Delta PL_f = -6\log_{10}(f/2000)$, where f denotes the frequency of the WiMAX network (given in MHz). The second correction term $\Delta PL_{h_{SS}}$ is set to $\Delta PL_{h_{SS}} = -10.8\log_{10}(h_{SS}/2)$ for urban and suburban terrain types, and to $\Delta PL_{h_{SS}} = -20\log_{10}(h_{SS}/2)$ for the rural terrain type, where h_{SS} denotes the height of the SS antenna (2 m $< h_{SS} <$ 10 m).

Next, we introduce the maximum QoS-coverage approach as a cost-modeling technique. The maximum QoS-coverage technique aims at maximizing the range of a WiMAX network with QoS support for different traffic types. In our cost-efficient network design, we consider the technological constraints of WiMAX networks. Based on the QoS requirements of the five different traffic classes of WiMAX, the number of QoS-aware SSs in range L of the BS is given by

$$n_{SS}(L) = \frac{Cap(L)}{C_{TC}}, \tag{13.10}$$

where $Cap(L)$ and C_{TC} denote the WiMAX network capacity in range L and the QoS requirements of a given traffic class TC, respectively.

The CAPEX (given in US $) of the designed WiMAX network with QoS support for SSs in range L is given by

$$C_{WiMAX}(L) = C_{BS} + n_{SS}(L) \cdot C_{SS}, \tag{13.11}$$

where C_{BS} and C_{SS} denote the equipment and installation costs of BS and SS, respectively.

Note that the power consumption of wireless equipment depends on their applied frequency ranges. Unlike previous studies, which considered only urban terrain (Baliga *et al.* [2008]), we take three different terrain types into account. As the power consumption of the BS depends on the requested average bandwidth of the attached SSs, the total power consumption of a WiMAX network (given in watt) for SSs located in range L is given by

$$P_{WiMAX}(L) = n_{SS}(L) \cdot \left(P_{SS} + 2 \cdot \frac{r_{SS}(L)}{R_{BS}(L)} \cdot P_{BS} \right), \tag{13.12}$$

where P_{SS} and P_{BS} denote the power consumption of each SS and BS, respectively. In Eq. (13.12), $R_{BS}(L)$ and $r_{SS}(L)$ denote the capacity of BS and requested average bandwidth of each SS in range L, respectively. We note that the factor 2 is used to account for the power consumed for both service provisioning and heat dissipation (Gladisch *et al.* [2008]).

Unlike EPON, WiMAX network failures are limited to BS and SS outages. The probability of a WiMAX network failure due to a single node failure or multiple node failures (i.e., BS or SS in range L) is given by

$$Pr_{F-W}(L) = p_{NE}^{f} \times \left(\frac{\binom{n_{SS}(L)}{f} + \binom{n_{SS}(L)}{f-1}}{\binom{n_{SS}(L)+1}{f}} \right), \tag{13.13}$$

where p_{NE} denotes the probability of a network element (i.e., BS or SS) failure and f denotes the number of node failures, whereby $f \geq 1$.

The cost of repairing a single or multiple failures in a WiMAX network is given by

$$C_{R-WiMAX} = \sum_{\forall i} [Pr_i(L) \times (\alpha \cdot CAPEX(i))], \tag{13.14}$$

where i denotes the failed device (which can be a BS or SS in range L), $Pr_i(L)$ denotes the probability of failure for equipment i, deduced from Eq. (13.13), $CAPEX(i)$ denotes the initial equipment and installation cost of device i, and α denotes the repairing or replacing factor, as defined in Eq. (13.5).

The OPEX (given in US $) of the designed WiMAX network is then given by

$$O_{WiMAX} = C(P_{WiMAX}(L)) + C_{R-WiMAX}, \tag{13.15}$$

where $C(P_{WiMAX}(L))$ denotes the cost of power consumed by the BS and SSs located in range L from the BS.

13.3 Numerical results

Table 13.2 shows typical costs for equipment and installation, as well as power consumption of commercially available EPON and WiMAX networks used in our simulations. We consider an EPON with different number of ONUs and set the maximum distance of a pair of adjacent ONUs in the three different terrains (l_{urban}, $l_{suburban}$, and l_{rural}) to 500, 1000, and 1500 meters, respectively. The bit error rate (BER) of the wireless channel is set to 10^{-9} and for all three terrain types the path losses h_{BS} and h_{SS} are set to 20 and 3 meters, respectively. Furthermore, the operating frequency f is set to 5000 MHz. The hierarchical scheduler proposed in (Ghazisaidi *et al.* [2009]) is used as a WiMAX scheduler. For the generation of voice traffic, we use the voice codec standard ITU-T G.711 where a packet of 160 bytes is generated every 20 ms without compression, translating into a constant bit rate (CBR) source rate of 64 kb/s. In addition, the fixed-size CBR voice packets contain 12, 8, and 20 bytes of real-time transport protocol (RTP), user datagram protocol (UDP), and Internet protocol (IP) headers, respectively. Further, we assume that there is no silence suppression. For the generation of video traffic, we deploy MPEG-4 to encode 600-byte packets at a data rate of 768 kbs/s which generates UDP CBR traffic, including 8 bytes and 20 bytes of UDP and IP headers,

Table 13.2. Typical EPON and WiMAX CAPEX & OPEX (given in US $).

	EPON				WiMAX	
	OLT	Splitter	ONU	Cabling	BS	SS
Equipment cost	2030	200	288	13 per km	49 000	350
Installation cost	600	120	120	7000 per km	7500	50
Power consumption	20 W	0	3.5 W	0	220 W	13 W

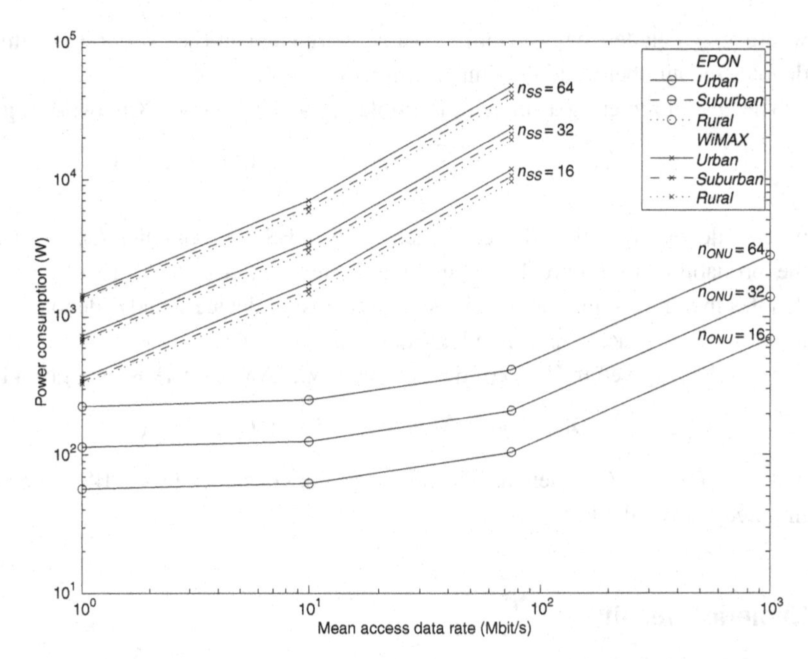

Figure 13.2 Power consumption vs. mean access data rate for EPON and WiMAX.

respectively. In our simulations, the voice and video codecs are used simultaneously, each encoding 50% of generated traffic. Moreover, we assume Poisson traffic with different packet sizes as background data traffic that uses 20% of the network capacity. The size of a generated data packet is equal to 40, 552, and 1500 bytes according to a distribution of 50%, 30%, and 20%, respectively. The generated data packets are transmitted with an additional 20-byte TCP header and 20-byte IP header.

Figure 13.2 shows the power consumption vs. mean access data rate for EPON and WiMAX serving $n_{ONU}, n_{SS} \in \{16, 32, 64\}$ ONUs/SSs at a range of 20 km for different terrain types. The power consumption increases for increasing mean access data rate, whereby EPON consumes less power than WiMAX. The power consumption of EPON is independent of the terrain type. The capacity of the BS in an urban setting is smaller than suburban and rural settings, resulting in an increased power consumption. For both EPON and WiMAX, the power consumption grows for an increasing number of ONUs and SSs.

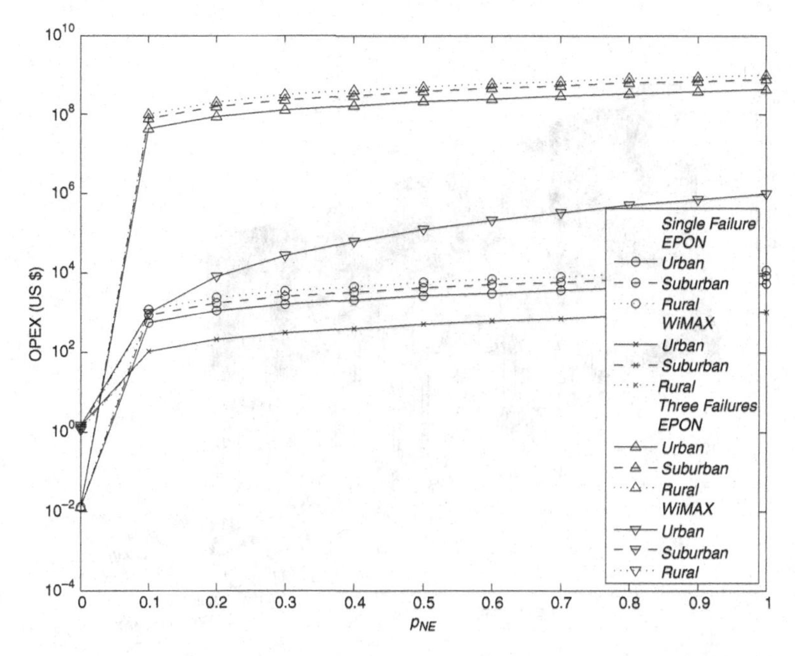

Figure 13.3 OPEX vs. network element failure probability p_{NE} for $n_{ONU} = n_{SS} = 32$ and a fixed mean access data rate of 75 Mb/s.

Figure 13.3 shows the OPEX vs. the network element failure probability p_{NE} for EPON and WiMAX with $n_{ONU} = n_{SS} = 32$ at a range of 20 km and a fixed mean access data rate of 75 Mb/s. The cost of power consumption is calculated for 0.0597 US $/kWh. The figure shows that the OPEX of EPON is less than that of WiMAX for the failure-free scenario (i.e., $p_{NE} = 0$). For a single failure, EPON OPEX is below WiMAX OPEX for $p_{NE} \leq 0.07$, and vice versa for higher p_{NE}. This observation holds also for three failures ($f_n = 2$, $f_f = 1$, and $f = 3$) whereby the OPEX curves of EPON and WiMAX already cross at $p_{NE} \approx 0.025$.

Figure 13.4 shows the total cost vs. range for EPON and WiMAX for $n_{ONU} = n_{SS} = 32$ with a single failure probability ($p_{NE} = 10^{-5}$) and a fixed mean access data rate of 75 Mb/s. Note that the maximum range of EPON is dependent on the terrain type and is determined by using the aforementioned maximum cost-sharing technique. The total cost of EPON increases for increasing range, while WiMAX total cost is largely independent of the range for a fixed number of SSs. We observe that WiMAX is more cost-efficient than EPON for a mean access data rate of up to 75 Mbit/s, especially for less populated suburban and rural terrain types. The cost difference between WiMAX and EPON becomes less pronounced for urban settings with high population densities. In fact, EPON might be a viable alternative to WiMAX in densely populated areas where the high installation costs of the required fiber infrastructure can be shared by a large number of subscribers. We note that the network revenue depends on the number of subscribers and the specific requirements for various traffic and services.

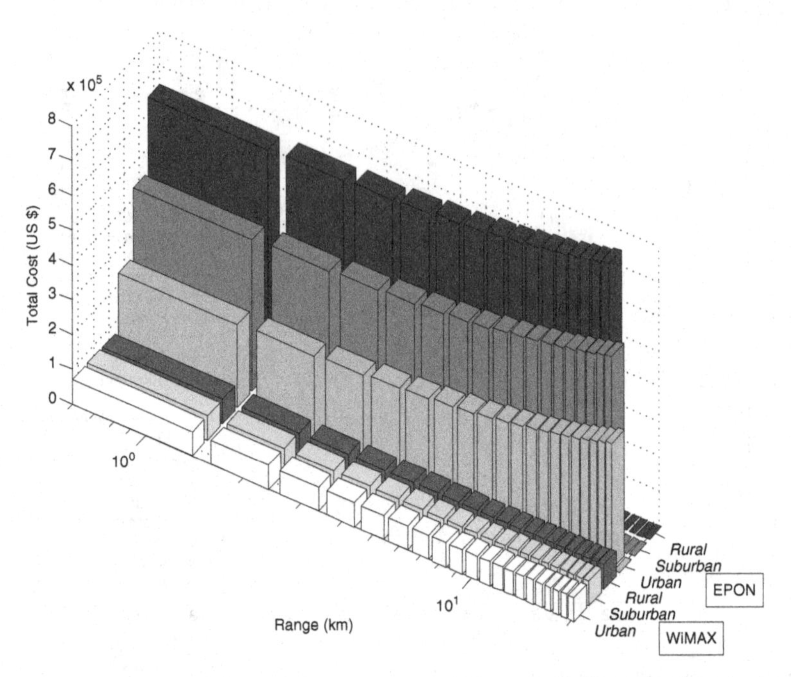

Figure 13.4 Total cost vs. range for $n_{ONU} = n_{SS} = 32$ and a fixed mean access data rate of 75 Mb/s.

While in WiMAX networks, each SS typically corresponds to a single end-user, it is important to note that an EPON ONU may serve multiple residential and/or business subscribers. Figure 13.5 shows the maximum number of subscribers (i.e., end-users) supported in EPON and WiMAX networks under voice, video, and triple-play traffic for urban terrain using the above-mentioned ITU-T G.711 and MPEG-4 codecs for the generation of voice and video traffic, respectively, and Poisson data traffic with the aforementioned trimodal packet size distribution. For all traffic types, EPON is able to support significantly more subscribers than WiMAX, thus providing a higher revenue potential to urban area network operators.

Finally, we compare the emerging next-generation of optical and wireless access networks (i.e., IEEE 802.3av 10 Gb/s EPON and IEEE 802.16m 1 Gb/s WiMAX) using our proposed techno-economic model. In our simulations, we conservatively estimate that the power consumption and cost of next-generation (NG) EPON and WiMAX equipment are twice as high as those of current EPON and WiMAX networks shown in Table 13.2, while installation costs are expected to remain unchanged. Figure 13.6 shows the power consumption vs. mean access data rate for next-generation EPON and WiMAX networks serving $n_{ONU}, n_{SS} \in \{16, 32, 64\}$ ONUs/SSs at a range of 20 km for different terrain types. Interestingly, the figure shows that the difference between next-generation EPON and WiMAX in terms of power consumption becomes less pronounced than in current EPON and WiMAX, especially at small mean access data rates (see Fig. 13.2 for comparison).

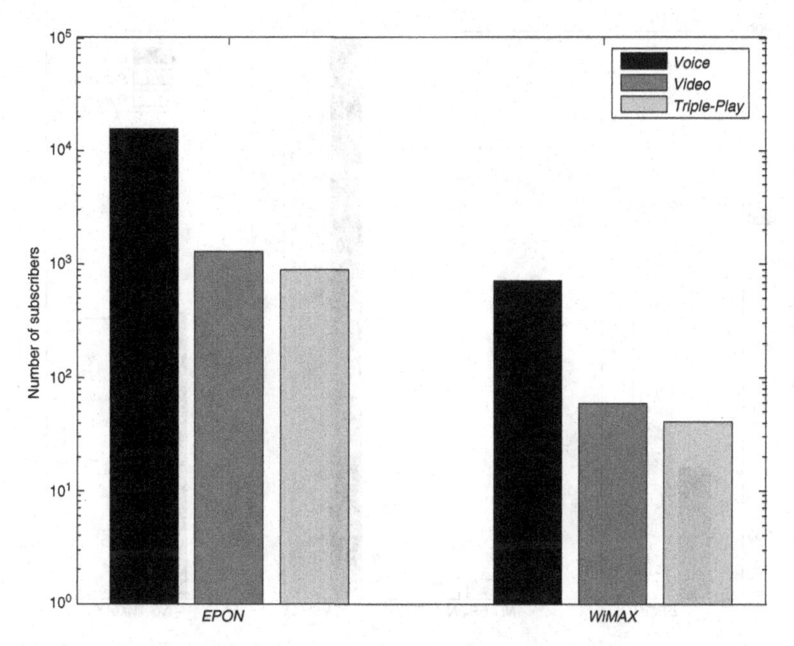

Figure 13.5 Number of subscribers in EPON and WiMAX networks under voice, video, and triple-play traffic for urban terrain.

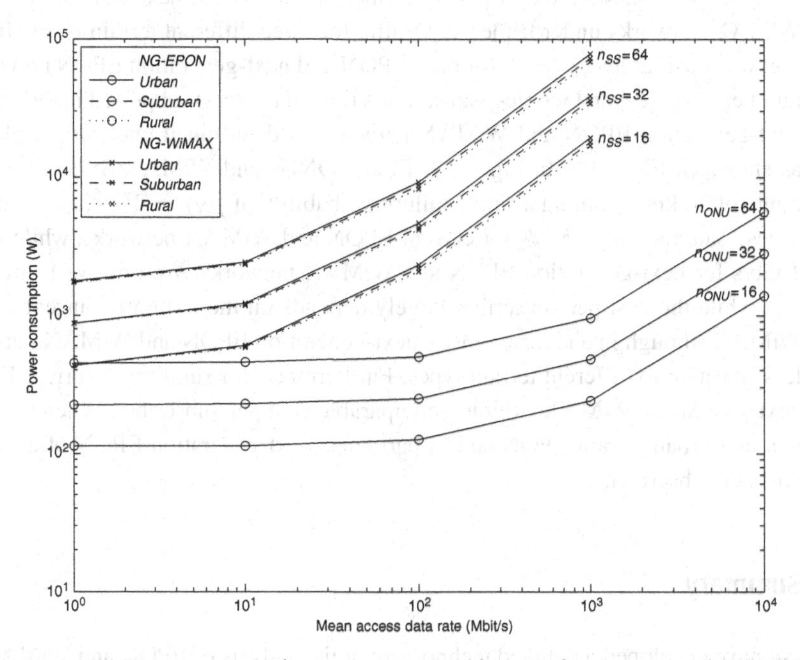

Figure 13.6 Power consumption vs. mean access data rate for next-generation (NG) EPON and WiMAX networks.

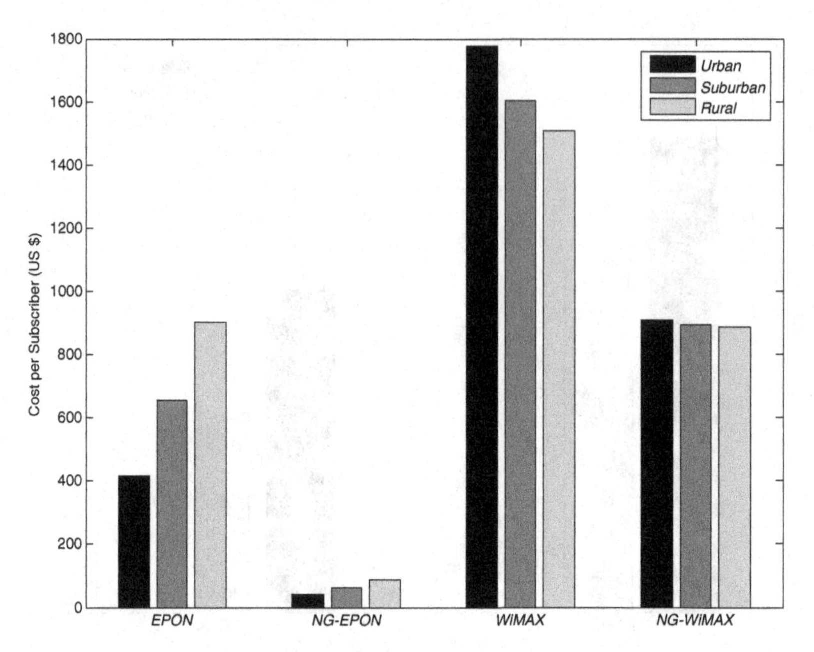

Figure 13.7 Cost per subscriber of current and next-generation (NG) EPON and WiMAX networks under triple-play traffic for three different terrain types.

Figure 13.7 shows the cost per subscriber of current and next-generation EPON and WiMAX networks under triple-play traffic for three different terrain types. In this figure, we consider $n_{ONU} = 32$ for both EPON and next-generation EPON networks. The number of optical and wireless subscribers (i.e., end-users) supported in the current and next-generation EPON and WiMAX networks is determined under triple-play traffic, as shown in Fig. 13.5. In Fig. 13.7, EPON ONUs and WiMAX SSs are located at a range of 20 km assuming a single failure probability of $p_{NE} = 10^{-5}$, and a fixed mean access data rate of 75 Mb/s for current EPON and WiMAX networks, while it is set to 1 Gb/s for next-generation EPON and WiMAX networks. We observe from Fig. 13.7 that while the cost per subscriber largely depends on the terrain in current EPON and WiMAX (though in a reverse sense), next-generation EPON and WiMAX networks are less sensitive to different terrain types. Furthermore, for rural areas current EPON and next-generation WiMAX exhibit a comparable cost per subscriber. Whereas for suburban and urban terrain EPON and in particular next-generation EPON offer the lowest cost per subscriber.

13.4 Summary

We have developed a detailed techno-economic analysis of EPON and WiMAX, which are key technologies in emerging FiWi broadband access networks. The obtained results show that the power consumption of EPON is smaller than that of WiMAX for different numbers of ONUs/SSs, terrain types, and mean access data rates. In terms of OPEX,

EPON is superior to WiMAX for deployment scenarios where network failures are less likely. Taking both OPEX and CAPEX into account, WiMAX is more cost-efficient than EPON, especially in suburban and rural areas with a small population density. For densely populated settings, EPON might be a viable alternative to WiMAX. In fact, EPON is able to provide data rates well above 75 Mb/s since, once put in place, EPON can be upgraded to much higher data rates by means of advanced time division multiplexing (TDM) and/or wavelength division multiplexing (WDM) technologies without requiring any modifications of the installed fiber infrastructure. In terms of power consumption, the difference between next-generation EPON and WiMAX becomes smaller than in current EPON and WiMAX networks.

14 Network coding

Among the several next-generation passive optical network (NG-PON) requirements are the provisioning of higher bandwidth per subscriber, an increased splitting ratio, and an extended maximum reach compared with current Ethernet PON (EPON) and Gigabit PON (GPON) architectures. NG-PONs may offer additional functionalities such as protection (to be discussed in detail in Chapter 15), support topologies other than conventional tree structures, and they enable the consolidation of access, backhaul, and metro network infrastructures (Lin [2008]). In addition, substantial research activity is currently focused on the convergence of optical and wireless access architectures into bimodal fiber-wireless (FiWi) access networks (Ghazisaidi and Maier [2011]), a key feature of NG-PONs.

An important goal of FiWi research is to combine the most promising technologies proposed for wireless and optical access. *Network coding (NC)* is an example of such technologies. Consisting of bit- or packet-level coding operations, NC has been shown to improve throughput, simplify routing, and provide robustness against transmission errors and failures in various packet networks (Ho and Lun [2008]). In a recent study, significant throughput gains were demonstrated experimentally in NC-enabled WiFi-based mesh networks (Katti *et al.* [2008]).

In this chapter, we study the integration of NC within NG-PONs. The aim is to illustrate the NG-PON architectures where NC yields potential performance gains. Our illustrations and simulations demonstrate significant potential performance improvements while clarifying some underlying topological constraints of NC in various NG-PON scenarios. We will first consider the case of network coding in a conventional PON. Subsequently, we define inter-flow and intra-flow NC and discuss their applicability to NG-PONs. We then demonstrate the potential of NC in different NG-PON settings though illustrative examples. For further details we refer the interested reader to (Fouli *et al.* [2011]).

14.1 Networking coding in PON

NC has only just started to be investigated in the context of PONs (Belzner and Haunstein [2009], Miller *et al.* [2009]). Figure 14.1 illustrates the potential of inter-flow NC to improve throughput in current PONs. In this illustrative scenario, two packets are exchanged between two optical network units (ONUs).

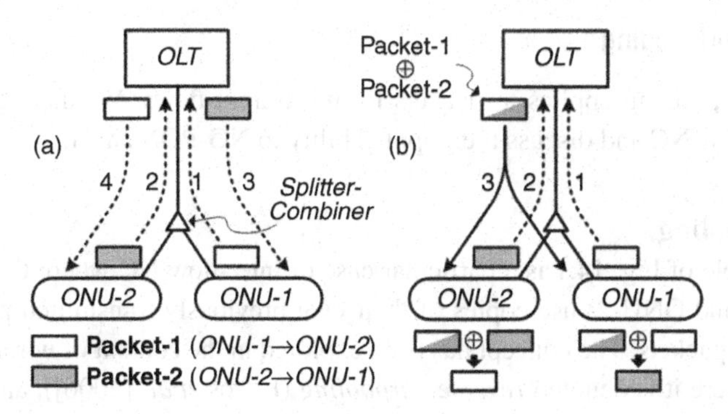

Figure 14.1 Network coding in a conventional passive optical network (PON).

Owing to the PON's directional splitter/combiner, ONUs may communicate only through the intermediary of the optical line terminal (OLT). In conventional PONs, such an exchange is usually performed in four separate packet transmissions, with the OLT receiving and then broadcasting each packet individually (see Fig. 14.1(a)). With NC, the OLT may code the received packets into a single packet using a simple bitwise exclusive-OR (XOR) operation, denoted by ⊕ (see Fig. 14.1(b)). Upon receiving the coded packet, the ONUs decode the packets destined to them using a copy of their previously transmitted packets. NC hence achieves the packet exchange in only three packet transmissions, using 50% less downstream bandwidth than conventional PONs. We will study this scenario in more detail below and provide compelling simulation results showing the impact of NC on PON performance.

14.2 Network coding in NG-PONs

NC stems from the observation that the function of nodes in a communications network is not restricted to routing, switching, and forwarding. In NC, nodes may perform operations on data units (e.g., bits, packets), generally using linear algebraic approaches, in order to improve network performance (Ho and Lun [2008]).

In the following, a flow is defined as a stream of data units with the same source and destination. Linear coding denotes the linear combination of individual symbols, defined over finite fields or vectors thereof, such that their extraction at the decoding node is possible through solving linear equations. Although nonlinear coding schemes are mentioned in the literature (Ho and Lun [2008]), we restrict the discussion in this chapter to the simpler and more practical linear coding. The coding of two packets or flows denotes the linear combination of their consecutive symbols using the same coefficients. Hence, any coded symbol, packet, or flow can be expressed as a linear combination $\sum c_i x_i$, where x_i denotes a native (i.e., uncoded) symbol, packet, or flow, and c_i represents the coding coefficient. In the case of a binary field, symbols consist of single bits, and both coding and decoding are performed through XOR operations (Katti *et al.* [2008]).

14.2.1 Inter-flow network coding

In inter-flow NC, coding applies to packets from different flows. We distinguish two forms of inter-flow NC and discuss their applicability to NG-PONs next.

Reverse carpooling

The PON example of Fig. 14.1 is a particular case of inter-flow NC where the receiver nodes (ONU-1 and ONU-2) use copies of their own previously transmitted packets to decode received packets. The concept has been explored in the context of wireless communications, where it is denoted *reverse carpooling* (Effros *et al.* [2006]), also known as piggybacking, or pairwise XOR coding. Reverse carpooling requires the uplink from the information-exchanging nodes towards a common intermediate node to be unicast while the downlink from the intermediate node back to the transmitting nodes must be broadcast. NC can hence exploit the underlying broadcast architecture to convert unicast transmissions into more efficient broadcast transmissions, as depicted in Fig. 14.1.

In wireless networks, (Katti *et al.* [2008]) uses reverse carpooling to increase throughput by exploiting the broadcast nature of wireless mesh networks. Each node is required to (1) store overheard packets that are not destined to it for a limited period of time, a procedure termed opportunistic listening, and (2) periodically send control packets called reception reports to inform its neighboring nodes of its stored packets. This enables nodes to *code opportunistically*: at each transmission, nodes combine the maximum number of packets that can be decoded at their next hop.

The example of Fig. 14.1 shows that the conditions for reverse carpooling are satisfied in conventional PONs, owing primarily to the use of the directional coupler. Furthermore, since only one intermediate node exists (i.e., the OLT), NC may be applied in a centralized manner. This removes the requirement for signaling (i.e., reception reports) and facilitates the integration of NC within PONs. Favorable conditions for reverse carpooling are pervasive in the NG-PONs that we consider since many of them use coupler-based tree architectures, including splitter-based WDM EPONs (McGarry *et al.* [2006]), LR-PONs (Shea and Mitchell [2007]), and integrated access-metro network architectures (Kazovsky *et al.* [2007], Maier *et al.* [2007]). However, reverse carpooling is not possible when connections with the intermediate node are reduced to point-to-point links such as in wavelength-routing WDM PONs. Similarly, when the medium is fully broadcast, inter-flow NC is not feasible for lack of intermediate nodes. In NG-PONs, this may occur when the nodes are connected through a reflective or star coupler.

Multipath inter-flow network coding

In multipath inter-flow NC, a receiver uses different linear combinations of the coded packets from different paths in order to successfully perform decoding. Multipath inter-flow NC is particularly relevant for multicasting, when flows are transmitted from multiple sources to multiple destinations across a shared network infrastructure where capacity bottlenecks arise.

Unlike reverse carpooling, multipath inter-flow NC requires multiple paths from the source to the destination. This renders it unfeasible in tree networks such as PONs, LR-PONs, and access-metro networks dominated by tree topologies, e.g., (Kazovsky *et al.* [2007]). Nevertheless, more diversified access-metro architectures and FiWi networks provide interesting possibilities, as shown in the examples below.

14.2.2 Intra-flow network coding

Rather than coding packets of different flows, *intra-flow coding* implies the coding of consecutive packets from the same flow and has been proposed in particular to improve reliability mechanisms in wireless networks (Fragouli *et al.* [2007]). As an alternative to acknowledgment-based repetition, a source node generates random linear combinations of the next N packets in the flow until N linearly independent ones are successfully received, enabling the destination node to decode the N native packets. Such *batch coding* does not require acknowledgment for each packet, but rather the entire batch, thus signaling the source to end the coded transmissions for that batch.

Different implementations of intra-flow NC have been examined and a comparison of their performances provided in (Lun *et al.* [2006]), among others. Intra-flow NC may be applied in an end-to-end fashion, similarly to fountain codes, or with encoding at intermediate nodes. The ability to re-encode at intermediate nodes is particularly important for dead spot mitigation and in multicasting scenarios (Fragouli *et al.* [2007]).

Coding may be applied along a sliding window rather than in fixed-size batches. In this scheme, decoding occurs as soon as the destination receives enough linear combinations for any subset of native packets. Both batch-based and sliding-window techniques deliver native packets to higher layers only after decoding events (i.e., the arrival of enough linearly independent packets to perform decoding). While one could expect this to have a possible adverse effect on delay-sensitive applications (e.g., voice, streaming), the overall delay required to transfer content over lossy links is reduced (Eryilmaz *et al.* [2006]).

In the context of NG-PONs, end-to-end NC may be applied between any source and destination to reduce the complexity of feedbacks and to increase reliability against packet losses due to link losses or congestion. In particular, end-to-end coding mechanisms may be implemented across the wireless part of a FiWi network to alleviate wireless link losses. Since they require the existence of multiple paths from source to destination, general coding methods cannot be deployed across tree-based NG-PONs. Nevertheless, they may be employed in metro ring networks and FiWi networks. In addition to increasing throughput and reducing delay in the presence of packet losses, they provide inherent reliability enhancement, as explained in the next section.

14.2.3 Metro-access networks

In contrast to tree-based networks such as PONs, integrated metro-access networks feature more opportunities for multiple paths between sources and destinations where NC may be applied. For example, Fig. 14.2 illustrates the use of multipath inter-flow NC for multicasting within hybrid ring-star metro networks. Note that this hybrid ring-star topology is a practical example of the "butterfly" configuration widely discussed in theoretical NC works (Ho and Lun [2008]).

The hybrid ring-star topology was shown to improve the resilience, spatial reuse, and bandwidth efficiency of packet-based optical metro rings (Herzog and Maier [2006]). Figure 14.2(a) illustrates such an architecture, where a subset of the ring nodes are attached to a single-hop WDM star network built from widely available metropolitan dark fiber, whereby the hub of the star subnetwork consists of a broadcast passive star coupler (PSC). Hybrid ring-star architectures are powerful metro ring candidates because they allow cautious WDM upgrades and exploit low-cost passive technology and dark fibers. In addition, they may be deployed to all-optically interconnect multiple TDM/WDM PONs (Maier *et al.* [2007]). Although different star network architectures were proposed in (Herzog and Maier [2006]), the PSC implementation is of particular interest here owing to its wavelength broadcasting nature. Using the PSC, star nodes such as node n can use a single transmission to broadcast packets to the star nodes.

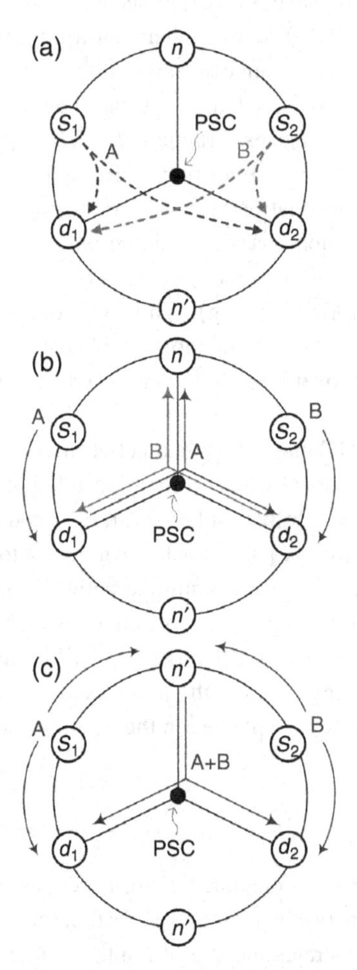

Figure 14.2 Inter-flow network coding in ring-star metro network.

We assume the multicast requests of Fig. 14.2(a): each of the sources s_1 and s_2 multicasts one flow to destinations d_1 and d_2 simultaneously. Each of the flows A and B, originating from s_1 and s_2, respectively, requires the capacity of a single wavelength. Assuming shortest path routing (i.e., minimum hop routing), Fig. 14.2(b) depicts a possible routing configuration, where each destination receives one flow over the ring and forwards it through the star subnetwork to the second destination. Note that the routing scheme of Fig. 14.2(b) requires two wavelengths on the star network. Furthermore, although other shortest paths exist, they all require two wavelengths on the star network.

In the NC solution (Fig. 14.2(c)), copies of the flows are routed through node n, where they are coded and broadcast through the star subnetwork. In this example of inter-flow coding, each destination receives one coded and one native flow through different paths and is thus able to perform decoding. The use of inter-flow NC removes the requirement for an additional wavelength on the star subnetwork, hence realizing a 50% throughput gain, at the expense of higher spatial utilization on the ring.

14.3 Network coding in FiWi access networks

14.3.1 Performance enhancement

Some of the most promising applications of NC relate to FiWi networks. We illustrate the potential of intra-flow NC in FiWi networks through the example of Fig. 14.3(a). Suppose a node in the optical portion of an NG-PON is multicasting a set of packets $\{p_1, p_2, p_3\}$ to two wireless nodes r_1 and r_2. The destination nodes are located in the vicinity of wireless local area network (WLAN) access points AP-1 and AP-2 such that r_1 is connected to AP-1 whereas r_2 may connect to AP-1 or AP-2, as shown in Fig. 14.3(a). First, the OLT broadcasts the batch of packets to both APs. In a conventional WLAN setting, native packets are transmitted by the AP-1 and AP-2 in sequence, and each packet is separately acknowledged by r_1 and r_2. Typically, r_2 selects the access point with the strongest signal for the transaction. With NC, the access points keep transmitting random linear combinations of the native packets without waiting for acknowledgments. Hence, AP-1 and AP-2 transmit the sequences $\{p'_1, p'_2, p'_3, \ldots\}$ and $\{p''_1, p''_2, p''_3, \ldots\}$, respectively (see Fig. 14.3(a)). Once r_1 and r_2 receive enough independent linear combinations to decode the native packets (i.e., three), they use a single block acknowledgment for the whole batch.

For illustration, the following simplifying assumptions were made:

- Time is slotted, packet transmission in both upstream and downstream directions takes one timeslot.
- The medium access control (MAC) protocol avoids interference by imposing the cyclic transmission pattern of Fig. 14.3(b), where the solid and dashed arrows indicate packet and acknowledgment transmissions, respectively.
- The wireless channel experiences losses in 25% of the timeslots.
- When NC is not used, access points wait for one timeslot before retransmitting a packet, unless an acknowledgment is received.

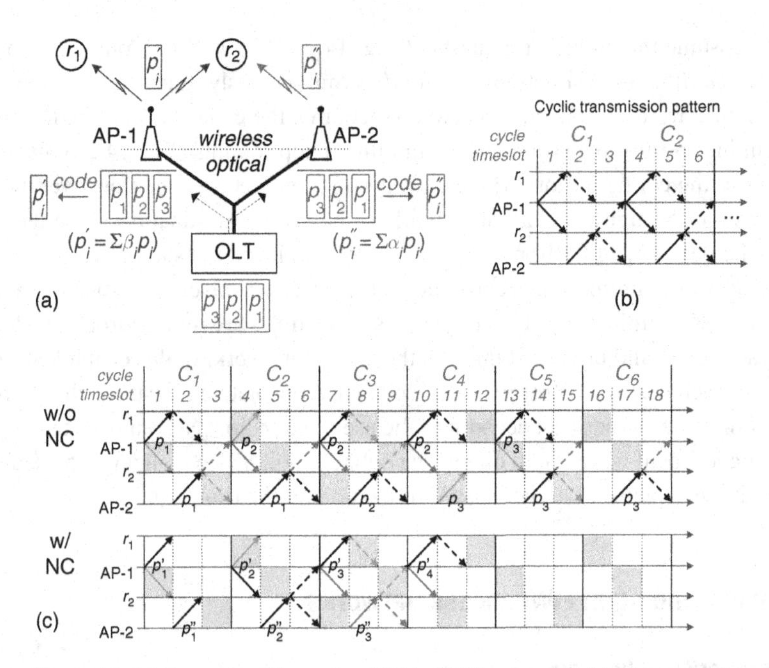

Figure 14.3 (a) Example of intra-flow NC in FiWi networks, (b) illustrative transmission pattern, and (c) time-space diagram.

The time-space diagrams of Fig. 14.3(c) illustrate the difference between coding (lower diagram) and no coding (upper diagram). The shaded squares represent time-slots with channel losses, where the same channel loss pattern is assumed with and without NC. When NC is not employed (upper diagram), r_1 receives all three packets at timeslot 13 and needs 9 packet transmissions for the transaction. In contrast, with intra-flow NC (lower diagram), r_1 is able to decode all three packets at timeslot 10 and uses a total of 6 transmissions, thus achieving gains of 23% and 33% in delay and energy, respectively.

Being connected to both access points simultaneously, r_2 achieves better performance. In conventional WLANs, r_2 selects the strongest of the two signals. Hence, with no coding (upper diagram), r_2 starts ignoring broadcasts from AP-1 after timeslot 1. r_2 receives all three packets at timeslot 14 and requires 11 packet transmissions. Using NC (lower diagram), r_2 can receive both coded flows simultaneously and use them to decode the native packets; r_1 is thus able to decode all three packets at timeslot 5 and uses 3 transmissions, thus achieving gains of 64% in delay and 72% in energy. Note that delay and energy gains are still significant if r_2 ignores AP-1, reaching 42% and 54%, respectively. Moreover, the use of two different paths to r_2 enhances the reliability of the transfer against failures along the wireless paths. Overall, the example of Fig. 14.3 shows that intra-flow NC reacts more efficiently to the losses of the wireless medium.

Alternatively, coding may be implemented within the optical part of the FiWi network. In this case, the coded packet streams $\{p_1', p_2', p_3', ...\}$ and $\{p_1'', p_2'', p_3'', ...\}$ are generated by the OLT and halted by an acknowledgment from the access points once the wireless nodes have received three independent linear combinations. Such a

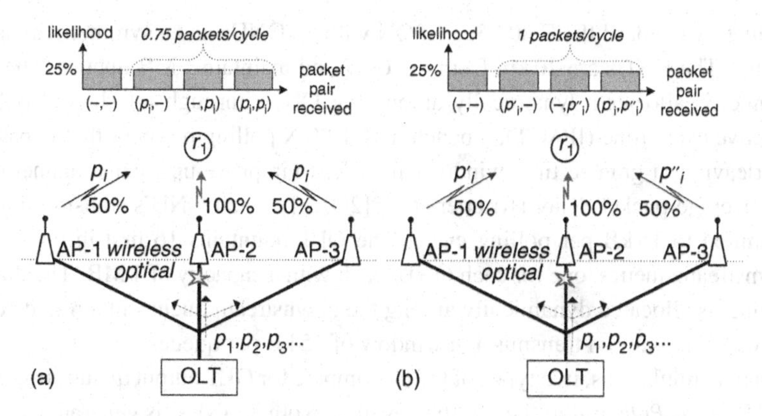

Figure 14.4 FiWi network survivability.

configuration leverages the higher optical bandwidth and processing capabilities. Furthermore, it is required when the PON is a point-to-point medium (e.g., WDM PON) rather than a broadcast medium.

14.3.2 Resilience

NC may also potentially improve the resilience of FiWi networks to fiber cuts, as shown in Fig. 14.4. Suppose a flow $\{p_1, p_2, p_3...\}$ is to be transmitted from the OLT to r_1. For each of the links between r_1 and the three access points in its vicinity, Fig. 14.4(a) shows the probability of success per packet transmission. For illustration purposes, we assume that time is divided equally among wireless nodes, so that each node is allowed to transmit one packet per transmission cycle. Also, let the flow be initially routed via AP-2 such that r_1 receives it at a rate of one packet per cycle, ignoring acknowledgments. Figure 14.4(a) shows a cut in the distribution fiber between the OLT and AP-2. Clearly, if r_1 picks a single replacement path via AP-1 or AP-2, the flow may only be delivered at an average rate of 0.5 packets per cycle.

Figures 14.4(a) and (b) depict two alternative solutions where both lossy links are used simultaneously as backup paths. In Fig. 14.4(a), each native packet p_i $(i = 1, 2, 3, ...)$ is transmitted from both AP-1 and AP-3. In Fig. 14.4(b), however, the random linear combinations p_i' and p_i'' are transmitted from AP-1 and AP-3, respectively. Assuming that losses across the two links are independent, Figs. 14.4(a) and (b) show the likelihoods of the packet pairs received by r_1 each cycle. Without NC (Fig. 14.4(a)), the average packet rate attained is 0.75 packets per cycle. NC, however, enables r_1 to receive the flow at its full rate of 1 packet per cycle (Fig. 14.4(b)).

14.4 Numerical results

In this section, we use examples or numerical simulations to illustrate some of the potential NC applications in NG-PONs. Figure 14.1 represents the generic framework for applying reverse carpooling to intra-PON unicast transmissions. In the following, we simulate a standard IEEE 802.3ah EPON with 16 ONUs and a symmetrical data rate of 1 Gb/s. The ONUs are 20 km from the OLT and maintain 1 MB queues. The upstream channel is allocated dynamically among the ONUs through interleaved polling with adaptive cycle time (IPACT), a benchmark EPON polling protocol that is based on the interleaving of granted time windows in order to improve upstream channel utilization and average packet delay (Kramer *et al.* [2002a]). Each ONU's transmission window is limited to 15 kB per polling cycle. The OLT maintains 16 first-in-first-out (FIFO) downstream queues, one for each ONU, each with a capacity of 1 MB. The downstream channel is allocated dynamically among the downstream queues in a round-robin fashion with a maximum transmission window of 15 kB per queue.

In our simulations, two types of traffic compete for OLT output queue space: (1) at the ONUs, *intra-PON* traffic (i.e., traffic destined to other ONUs) is generated for upstream transmission, and (2) at the OLT, an *external* traffic stream of packets destined to the ONUs is injected representing traffic generated outside the EPON. We assume Poisson traffic, where the packet size is uniformly distributed over the Ethernet packet size range (64B–1518B). In addition, the destination of both intra-PON and external packets follows a uniform distribution over all ONUs. After a 5 s warmup period, we simulate the transmission of 10^5 packets.

Opportunistic coding is integrated within layer-2 as follows. Each intra-PON packet to be transmitted downstream and having source ONU-i and destination ONU-j is coded with the earliest packet having inverted source and destination (i.e., with source ONU-j and destination ONU-i). If no such packet exists at the time of transmission, the packet is transmitted uncoded. To determine the effects of NC, we fix the external traffic rate to 0.5 Gb/s and vary the intra-PON traffic rate from 0.1 Gb/s to 0.9 Gb/s. Figure 14.5 compares the performance of native and NC-enhanced EPON in terms of mean aggregate throughput (Fig. 14.5(a)), average OLT downstream queue size (Fig. 14.5(b)), and mean delay (Fig. 14.5(c)). In Fig. 14.5, native and NC-enhanced EPON are represented through dashed and solid plots, respectively, with 95% confidence intervals.

The aggregate throughput plots of Fig. 14.5(a) show that coding gains appear at the point of congestion, when the intra-PON traffic load is 0.5 Gb/s. This point corresponds to the input aggregate traffic level (of both intra-PON and external packet streams) reaching the downstream data rate. As the OLT downstream queues grow, more coding opportunities arise, and the coding gain increases almost to 30% (0.2 Gb/s) for the intra-PON traffic. It is important to note that throughput gains are also achieved by the uncoded external traffic stream, reaching 27% (0.1 Gb/s) at the highest intra-PON traffic load.

To shed more light on the throughput gains, we turn to the average queue size plots of Fig. 14.5(b). Figure 14.5(b) represents the average steady-state size of all OLT downstream queues. On the one hand, the queue in native EPON expectedly saturates when aggregate downstream traffic rates exceed the data rate (intra-PON load of 0.6 Gb/s), translating into the loss of all excess packets, and the flattening of the throughput curve. However, this is not the case when NC is employed, as the queue remains two orders of magnitude below its saturation level, thus avoiding any significant packet losses and allowing the throughput to continue rising. The downstream queues eventually saturate

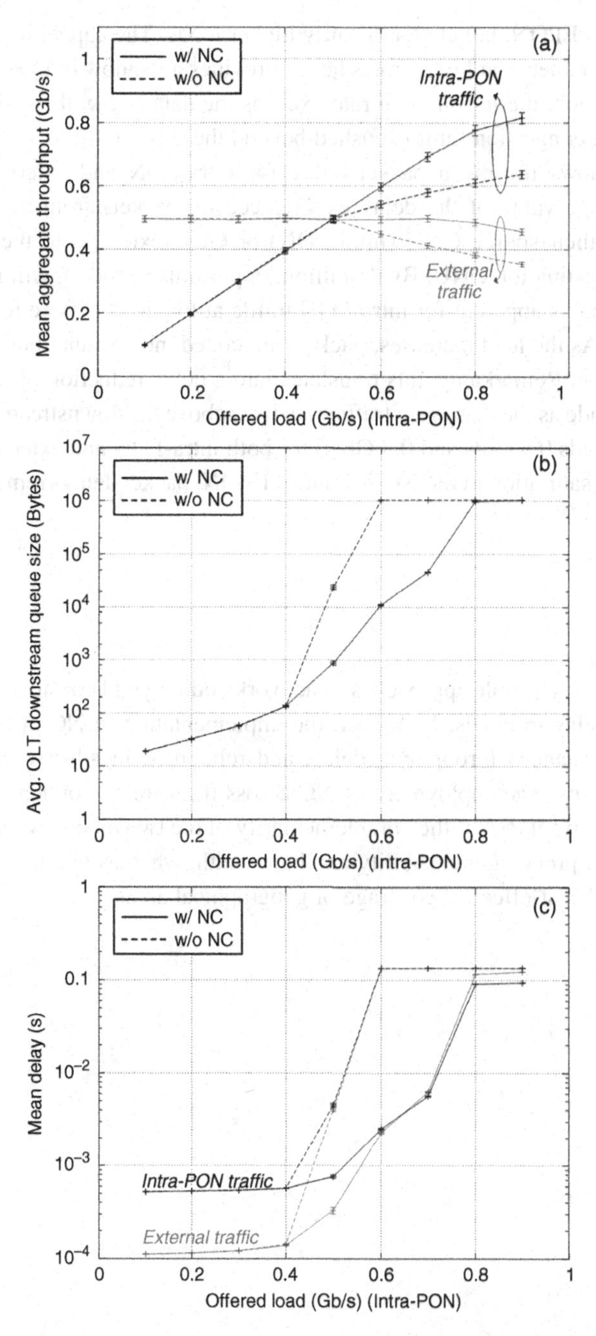

Figure 14.5 Performance enhancement through network coding (NC) in EPON: for a constant external traffic load (0.5 Gb/s) and increasing intra-PON traffic load values, we plot (a) mean aggregate throughput, (b) average OLT downstream queue size, and (c) mean delay. The solid and dashed curves are plotted with and without NC, respectively. The results in (a) and (c) are shown for intra-PON (black) and external (grey) traffic.

for the NC-enhanced EPON, but at significantly higher loads. The capability of NC to drain the downstream queues at higher rates hence provides a window of operation (0.5 Gb/s to 0.8 Gb/s), where the information rate exceeds the data rate without significant losses and where the congestion limit is pushed beyond the capacity limit.

Figure 14.5(c) shows the mean packet delay for intra-PON and external traffic, defined as the average value of the delay experienced by packets from the moment they are queued at their source ONU (intra-PON) or OLT (external) to the moment they arrive at their destination ONU. By definition, opportunistic coding will introduce no delay penalty. This is apparent for intra-PON traffic at low loads where few coding opportunities exist. As the load increases, packets are coded more often, thus spending less time in the queue. Remarkably, this translates into a delay reduction of more than one order of magnitude as the aggregate traffic rate rises above the downstream data rate (intra-PON traffic loads 0.6 Gb/s and 0.7 Gb/s) for both intra-PON and external traffic. As queues approach saturation in the NC-enhanced EPON, packet delays remain below native EPON levels.

14.5 Summary

We have shown that very simple approaches to network coding yield considerable gains in throughput and delay in PONs. Moreover, the implementation of NC in NG-PONs holds promise for enhanced throughput, delay, and reliability, in adverse conditions with high packet losses. The deployment of NC across the wireless–optical boundary may reap particular advantages of the complementarity of the two media, where optical networks provide the processing capability and bandwidth, whereas wireless networks provide mobility and cost-effective coverage of geographical areas.

15 Optical and wireless protection

Fiber-to-the-home (FTTH) or close to it (FTTx) networks are poised to become the next major success story for optical fiber communications. Future FTTx access networks unleash the economic potential and societal benefit by opening up the first/last mile bandwidth bottleneck between bandwidth-hungry end-users and high-speed backbone networks. Owing to their longevity, low attenuation, and huge bandwidth, passive optical networks (PONs) are widely deployed to realize cost-effective FTTx access networks. Fiber has been envisioned for delivering broadband services for over 30 years. However, many roadblocks related to component and installation costs have slowed down the progress toward FTTx since it was first proposed. Currently, FTTH is being installed in many countries, but it still represents only a fraction of all deployed broadband lines (Shumate [2008]). The two major state-of-the-art PON standards IEEE 802.3ah Ethernet PON (EPON) and ITU-T G.984 Gigabit PON (GPON) consist both of a single upstream wavelength channel and a separate single downstream wavelength channel, whereby both channels are operated using time division multiplexing (TDM). EPON and GPON are expected to coexist for the foreseeable future as they evolve into next-generation PONs (NG-PONs) (Effenberger *et al.* [2007], Kazovsky *et al.* [2007]).

NG-PONs can be categorized into high-speed TDM PON, wavelength division multiplexing (WDM) PON, and long-reach PON (LR-PON) (Lin [2008], Effenberger *et al.* [2009a]). While current PONs are able to provide service only for a maximum of 256 optical network units (ONUs) located at a range of 20 km from the optical line terminal (OLT), LR-PONs are designed for longer distances between the OLT and ONUs of up to 100 km as well as larger numbers of ONUs (2000 to 4000 ONUs) (Talli and Townsend [2006]). Owing to the higher data rates, wavelength channel counts, number of ONUs, and coverage of NG-PONs, network survivability is becoming a key issue. While in conventional Gb/s PONs with typically 32–64 attached ONUs fiber link cuts only affect a relatively small number of subscribers, survivability of high-capacity NG-PONs with fibers carrying multiple wavelength channels in both upstream and downstream directions, each wavelength operating at 10 Gb/s, becomes increasingly important due to the fact that fiber link failures would result in the loss of significantly higher traffic volumes. NG-PONs such as high-speed 10 Gb/s PONs offer enough capacity not only for (best-effort) residential use but also for business applications, which require carrier-class survivability for business continuity (Mukai *et al.* [2009]).

The following techniques might be applied to resolve the survivability issues of NG-PONs:

1. **Full/partial PON duplication (optical protection):** With full PON duplication, a spare set of OLT and ONUs as well as optical fiber links and photonic network elements, e.g., splitters, is deployed to protect the primary optical infrastructure with a secondary backup (Koonen [2006]). While full PON duplication techniques in most cases might be cost-prohibitive, partial PON protection seems more viable, where only a subset of fibers are protected through stand-by fibers.
2. **Wireless mesh network (WMN)-based fiber-wireless (FiWi) architecture (wireless protection):** In a WMN-based FiWi network, a subset or all ONUs of an NG-PON network are equipped with wireless devices, referred to as mesh portal points (MPPs) in IEEE 802.11s. In the resultant WMN-based FiWi protection technique, an MPP forwards frames to another ONU via wireless links instead of going through the OLT. In an IEEE 802.11s based WMN, MPPs are able to send frames to each other directly (single-hop) or through intermediate wireless mesh points (MPs) by means of multi hopping. We note that WMN can be deployed using any wireless local area network (WLAN) technologies, such as IEEE standards 802.11 a/b/g/n or the emerging very high throughput (VHT) WLAN, whereby the maximum distance between two adjacent WMN nodes (i.e., maximum length of one wireless hop) depends on the applied wireless technology. For instance, the maximum distance between two adjacent WMN nodes has to be less than 2.7 km (line-of-sight (LOS)) in IEEE 802.11n for a slot size of 9 μs and short interframe space (SIFS) of 16 μs. Applying orthogonal frequency division multiplexing (OFDM) and multiple input multiple output (MIMO) antennas in the physical (PHY) layer of IEEE 802.11n WLANs provides various capabilities, such as antenna diversity (selection) and spatial multiplexing. The use of multiple antennas in next-generation WLAN-based WMNs provides multipath capability, which increases both throughput and transmission range. In a fading channel, multiple antennas can increase the system reliability through spatial diversity. It was shown in (Lin and Wong [2008]) that by transferring the same data across different paths, multiple independently faded data symbols can be successfully received at the destination node and the transmission reliability is increased significantly. Furthermore, the enhanced PHY layer of next-generation WLANs applies two powerful adaptive coding schemes: space time block coding (STBC) and low density parity check coding (LDPC). According to (Vermesan *et al.* [2010], Eldin *et al.* [2009]), the robustness and reliability of next-generation WLAN-based networks can be improved significantly by using the two aforementioned coding schemes. In WMNs, multiple redundant paths exist throughout the network. Mesh nodes may re-route traffic along alternative paths. It was shown in (Egeland and Engelstad [2009]) that adding redundant mesh points to WMNs improves the wireless link availability and reliability significantly. Due to the nature of WMNs, their channels are not error-free. WLAN-based WMNs apply a data link layer error control technique, known as automatic repeat request (ARQ), which uses acknowledgment messages and packet retransmissions to achieve reliable data transfer. To further improve the network reliability, error correction schemes such as forward error correction (FEC) can be used. Hybrid adaptive FEC and ARQ techniques were proposed in (Wu *et al.* [2009]) to increase the reliability

of WLAN-based networks. We note that current IEEE 802.3ah EPON and IEEE 802.3av 10G-EPON provide 1 Gb/s and 10 Gb/s, respectively. On the other hand, next-generation WLAN-based WMN and emerging VHT WLAN-based WMN technologies are able to provide raw data rates of 600 Mb/s and 1 Gb/s, respectively. It is important to note that the bandwidth of EPON and PONs in general is shared among all ONUs. That is, under the assumption of bandwidth fairness, the data rate available to a single EPON ONU is equal to 1/32 Gb/s $\approx$ 31 Mb/s for a typical scenario of 32 ONUs, which is below the data rate offered by an 802.11n WLAN access point (AP) possibly attached to the ONU. Also note that unlike EPON (and PONs in general) WMNs allow for spatial reuse of bandwidth, whereby a given channel can be used multiple times in different regions of the WMN that do not spectrally overlap. As a result, the aggregate capacity of a WMN is beyond that of a single WLAN link.

Several recent studies have started to explore the benefits of using the wireless network resources to render FiWi access networks survivable against fiber link failures (Feng and Ruan [2009], Yubin *et al.* [2009], Schütz and Correia [2009], Correia *et al.* [2009], Kantarci and Mouftah [2010]). In this chapter, we analyze the survivability of optical NG-PONs and bimodal FiWi access networks and examine the benefit of upgrading selected ONUs with wireless equipment in order to improve network survivability in a pay-as-you-grow manner. In the following, we first evaluate the survivability of NG-PONs and FiWi networks by means of probabilistic analysis taking both optical and wireless protection into account. Second, we propose and examine different ONU selection schemes to improve the survivability of NG-PONs by means of wireless extensions and partial optical protection. Third, in our numerical work we study the impact of different network topologies on the survivability of NG-PON and FiWi networks for a wide range of fiber link failure scenarios. For further details we refer the interested reader to (Ghazisaidi *et al.* [2011]).

15.1 Survivability analysis

In this section, we analyze the survivability of NG-PONs and FiWi networks with and without partial optical protection. The considered LR-PON may be upgraded with WDM and/or high-speed transceivers to create WDM and/or high-speed TDM LR-PONs, respectively.

15.1.1 NG-PON without protection

Let us first consider an LR-PON tree-and-branch topology without any protection and investigate its connectivity as a function of different fiber link failure probabilities. In a multi-stage tree-and-branch based LR-PON with N connected ONUs, we denote the probability that a given fiber link fails by p_n, where n is the index of the stage to which the fiber link belongs, as shown in Fig. 15.1. Note that the special case of only two stages $n = 0$ and $n = 1$ would correspond to conventional TDM/WDM PONs with a

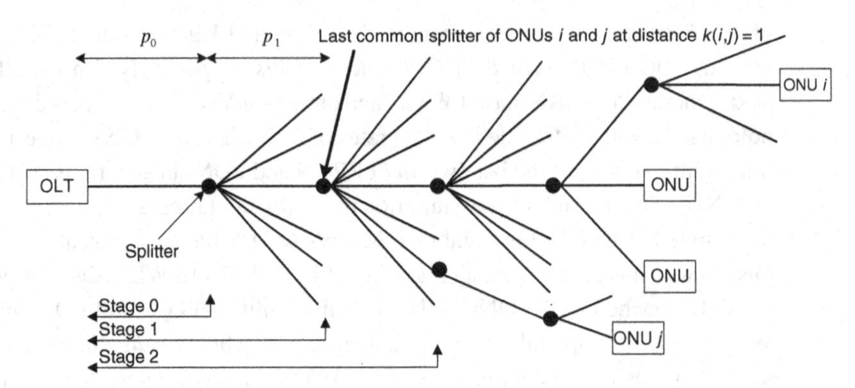

Figure 15.1 LR-PON and last common splitter of ONU i and ONU j for $d_i = 5$, $d_j = 4$, and $k(i, j) = 1$.

single splitter. Let $p_0, p_1, \ldots$ denote the fiber link failure probability of stage $0, 1, \ldots,$ respectively. We assume that fiber link failures occur independently from each other. Further, we define d_i as the distance (i.e., number of intermediate splitters) between a given ONU i and the OLT given by

$$d_i = distance(\text{ONU } i, \text{OLT}), i \in \{1, 2, \ldots, N\}. \tag{15.1}$$

Note that our model can be easily generalized by assuming different failure probabilities of fiber links belonging to the same stage (i.e., having the same distance to the OLT).

Let $k(i, j)$ be the distance of the last common splitter of ONUs i and j to the OLT (see Fig. 15.1 for illustration). The probability p_{ij} that ONUs i and j are connected via the OLT, which is equivalent to the probability that all fiber links from ONU i and ONU j to the OLT are intact, is given by

$$\begin{aligned} p_{ij} &= (1 - p_0) \cdots (1 - p_{k(i,j)}) \\ &\cdot (1 - p_{k(i,j)+1})^2 \cdots (1 - p_{d_i \wedge d_j})^2 \\ &\cdot (1 - p_{(d_i \wedge d_j)+1}) \cdots (1 - p_{d_i \vee d_j}), \end{aligned} \tag{15.2}$$

where $\wedge$ and $\vee$ denote the minimum and maximum values of two variables, respectively. In the first line of Eq. (15.2), the probability of all intact shared fiber links from the OLT to the last common splitter of the two ONUs is calculated (i.e., from p_0 to $p_{k(i,j)}$). In the second line of this equation, the probability of all intact fiber links from the last common splitter of the two ONUs to the ONU that has the minimum distance to the OLT (i.e., $d_i \wedge d_j$) is calculated. We note that in this calculation the ONUs apply different independent fiber links with the same failure probability, which results in the squared probabilities. The probability of all intact fiber links from the ONU with minimum distance (i.e., $(d_i \wedge d_j) + 1$) to the ONU with maximum distance (i.e., $d_i \vee d_j$) is calculated in the third line of Eq. (15.2). Further, the probability that ONU i is connected to the OLT is given by

$$q_i = (1 - p_0) \ldots (1 - p_{d_i}). \tag{15.3}$$

15.1.2 FiWi: NG-PON with wireless protection

To provide survivability against fiber link failures, we have to find a set $\mathcal{W}$ of ONUs and equip each of them with an MPP such that they can communicate wirelessly even if the LR-PON fiber infrastructure fails completely. Generally speaking, $\mathcal{W}$ should be chosen as small as possible while at the same time guaranteeing a high degree of survivability. To allow for pay-as-you-grow wireless upgrades of LR-PONs and satisfy given cost constraints, in general only a subset of ONUs are equipped with an MPP, i.e., we fix the cardinality of $\mathcal{W}$ to $|\mathcal{W}| = M \leq N$.

We propose the following selection schemes to identify the M ONUs and equip each of them with an MPP:

1. **Random selection:** In this scheme, M ONUs are randomly selected among the N ONUs.
2. **Uniform selection:** In this approach, the M selected ONUs include ONU 1 and ONU N. The other $M-2$ ONUs are uniformly selected among the remaining $N-2$ ONUs such that the index of two neighboring selected ONUs differs by $\lfloor N/(M-1) \rfloor$, i.e., $\mathcal{W} = \{\text{ONU } i, i = 1, 1 + \lfloor N/(M-1) \rfloor, \ldots, N\}$.
3. **Selection of weakest ONUs:** In this scheme, the M ONUs with the smallest probability q_i of being (optically) connected to the OLT are selected.
4. **Selection of strongest ONUs:** Conversely, this scheme selects the M ONUs with the largest probability q_i of being (optically) connected to the OLT.

Next, let $\mathcal{O}$ be the random subset of ONUs that are connected to the OLT optically after one or more fiber link failures have occurred. For a given set $\mathcal{W}$, the following two cases can happen:

- If $\mathcal{O} \cap \mathcal{W} = \varnothing$ (i.e., both sets are disjoint), then all ONUs in $\mathcal{O}$ and all ONUs in $\mathcal{W}$ can communicate among themselves, but no ONU in $\mathcal{O}$ can communicate with any ONU in $\mathcal{W}$.
- If $\mathcal{O} \cap \mathcal{W} \neq \varnothing$, then any pair of ONUs, say, ONUs i and j, in $\mathcal{O} \cup \mathcal{W}$ can communicate with each other even if ONU $i \in \mathcal{O}$ but $i \notin \mathcal{W}$ and ONU $j \in \mathcal{W}$ but $j \notin \mathcal{O}$, and vice versa.

For a given set $\mathcal{W}$, the expected number of ONUs, which are connected to the OLT is given by

$$\sum_{i=1}^{N} r_i, \tag{15.4}$$

where r_i is the probability that ONU i has a connection to the OLT, either directly optically or wirelessly optically, after equipping the selected M ONUs with MPPs. In Eq. (15.4), r_i is given by

$$r_i = \begin{cases} \mathbb{P}(\mathcal{O} \cap \mathcal{W} \neq \varnothing), i \in \mathcal{W} \\ \mathbb{P}(i \in \mathcal{O}) = q_i, i \notin \mathcal{W}, \end{cases} \tag{15.5}$$

whereby $\mathbb{P}(\mathcal{O} \cap \mathcal{W} \neq \phi)$ is the probability that there exists at least one ONU in $\mathcal{W}$ which is also in $\mathcal{O}$, i.e., there is at least one wirelessly upgraded ONU that has a failure-free optical connection to the OLT. In the following, we describe the computer program we used to recursively calculate $\mathbb{P}(\mathcal{O} \cap \mathcal{W} \neq \varnothing)$ according to the four following steps:

1. First, we prune the LR-PON tree-and-branch topology by removing all ONUs that are not in $\mathcal{W}$ and the branches leading to them. The resultant pruned tree topology contains only the M selected ONUs in $\mathcal{W}$, including only the branches connecting them to the OLT.
2. In the pruned tree topology, splitters with one incoming fiber link of stage i and one outgoing fiber link of stage $i+1$ are replaced with a single fiber link whose assigned failure probability is equal to $p_e = 1 - (1 - p_i)(1 - p_{i+1})$. That is, 1:1 splitters together with their respective incoming fiber link and outgoing fiber link are replaced with a new fiber link of failure probability p_e.
3. Beginning at the leaves (i.e., M ONUs), we assign each of the splitters that are directly connected to the M ONUs the probability that it has at least one failure-free optical connection to one of its corresponding ONUs. For a given splitter s_n, n hops away from the OLT, this probability is given by

$$\mathbb{P}_{s_n} = 1 - \prod_{i=1}^{m} (1 - p_{s_n i}), \tag{15.6}$$

   where m and $p_{s_n i}$ denote the number of ONUs connected to splitter s_n and the probability that ONU i is connected to splitter s_n, respectively. This step is repeated for all splitters of each stage of the pruned tree until we reach the first splitter s_1 next to the OLT.
4. Finally, we calculate the probability that the OLT has a failure-free optical connection to at least one of the M ONUs, which is given by

$$\mathbb{P}_{OLT} = (1 - p_O) \cdot \mathbb{P}_{s_1}. \tag{15.7}$$

   Note that $\mathbb{P}_{OLT}$ is equivalent to probability r_i in Eq. (15.5) for $i \in \mathcal{W}$. Thus, we have $\mathbb{P}(\mathcal{O} \cap \mathcal{W} \neq \varnothing) = \mathbb{P}_{OLT}$ for ONU i, $i \in \mathcal{W}$.

15.1.3 FiWi: NG-PON with both wireless and optical protection

So far, we have assumed that the LR-PON under consideration has no optical protection. Recall from above that partial optical protection is a viable solution to improve the survivability of LR-PONs by connecting one or more ONUs with additional back-up fiber links to the OLT. Now, if there exists an ONU, say, ONU 1, with a safe optical connection to the OLT (i.e., $q_1 = 1$) by means of optical protection and ONU $1 \in \mathcal{W}$, then $\mathbb{P}(\mathcal{O} \cap \mathcal{W} \neq \varnothing) = 1$ and $\sum_{i=1}^{N} r_i$ is maximal among all choices of $\mathcal{W}$ for a given $|\mathcal{W}| = M$, if we upgrade the $M - 1$ weakest ONUs (i.e., those ONUs with the lowest q_i) and ONU 1 with MPPs. Note that by maximizing $\sum_{i=1}^{N} r_i$ the mean number of ONUs connected to the OLT becomes maximal.

15.1.4 Failure-free connections among ONUs

To evaluate and compare the aforementioned selection schemes, we compute the average number of failure-free connections among ONUs (i.e., pairs of ONUs connected by optical and/or wireless links). The average number D of failure-free connections among ONUs is given by

$$D := \mathbb{E}\left[|\mathcal{O} \cup \mathcal{W}| \cdot (|\mathcal{O} \cup \mathcal{W}| - 1)\right]. \tag{15.8}$$

To compute D, we can write

$$D = \mathbb{E}\left[\left(\sum_{i=1}^{N} \mathbb{1}_{\mathcal{O}\cup\mathcal{W}}(i)\right)\left(\left(\sum_{j=1}^{N} \mathbb{1}_{\mathcal{O}\cup\mathcal{W}}(j)\right) - 1\right)\right], \tag{15.9}$$

where $\mathbb{1}_{\mathcal{O}\cup\mathcal{W}}(i)$ denotes the indicator function of subset $\mathcal{O} \cup \mathcal{W}$ for a given ONU i and is given by

$$\mathbb{1}_{\mathcal{O}\cup\mathcal{W}}(i) = \begin{cases} 1, i \in (\mathcal{O} \cup \mathcal{W}) \\ 0, i \notin (\mathcal{O} \cup \mathcal{W}). \end{cases} \tag{15.10}$$

Using this definition of the indicator function, we can extend Eq. (15.9) to

$$D = \sum_{i=1}^{N}\sum_{j=1}^{N} \mathbb{P}(i \in (\mathcal{O} \cup \mathcal{W}), j \in (\mathcal{O} \cup \mathcal{W})) - \sum_{i=1}^{N} \mathbb{P}(i \in (\mathcal{O} \cup \mathcal{W})), \tag{15.11}$$

where the first term is computed by distinguishing the following cases

$$\mathbb{P}(i \in (\mathcal{O} \cup \mathcal{W}), j \in (\mathcal{O} \cup \mathcal{W})) = \begin{cases} 1, i \in \mathcal{W}, j \in \mathcal{W} \\ p_{ij}, i \notin \mathcal{W}, j \notin \mathcal{W} \\ q_j, i \in \mathcal{W}, j \notin \mathcal{W} \\ q_i, i \notin \mathcal{W}, j \in \mathcal{W} \end{cases} \tag{15.12}$$

and the second term is given by

$$\mathbb{P}(i \in (\mathcal{O} \cup \mathcal{W})) = \begin{cases} 1, i \in \mathcal{W} \\ q_i, i \notin \mathcal{W}. \end{cases} \tag{15.13}$$

15.2 Numerical results

To facilitate a better understanding of survivability in NG-PONs, we first consider the impact of the number of stages and number of ONUs on the optical fiber connection of ONUs to the OLT without taking the WMN into account. Let us start with a conventional 2-stage PON and consider a typical number of $N = 16$ ONUs connected to the OLT. We then increase the number of stages and number of ONUs to form NG-PONs. More specifically, we consider a binary tree based PON, where each additional stage increases the number of ONUs by a factor of 2. Toward this end, we replace the 1:16 splitter of the conventional PON with a 1:2 splitter and attach two 1:16 splitters to the leaves of the 1:2 splitter. As a result, the new PON has three stages and supports 32 ONUs

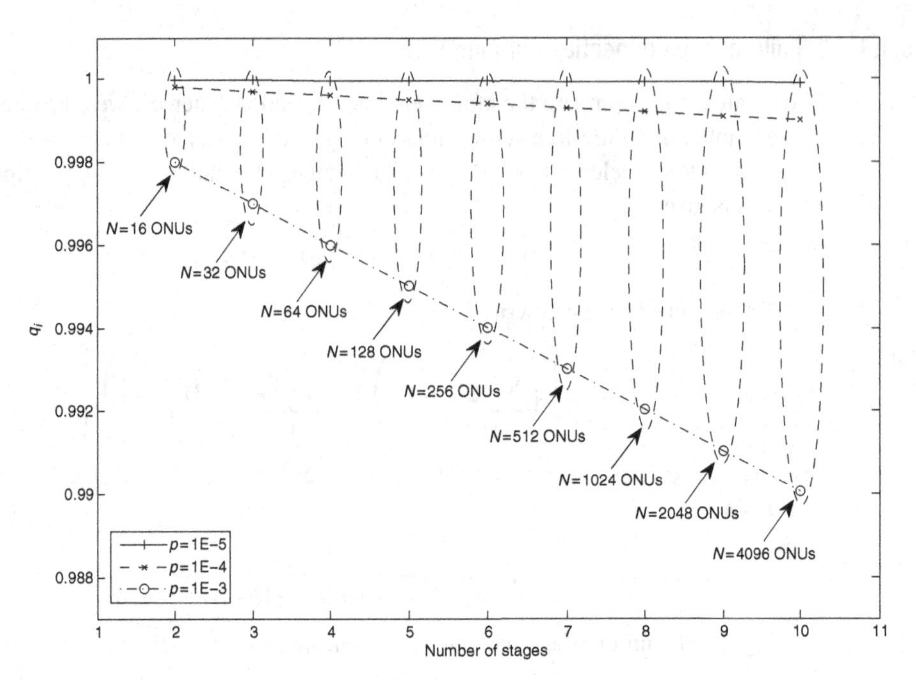

Figure 15.2 Impact of number of stages and number of ONUs on the probability q_i of an intact optical connection of ONU i to the OLT.

at a distance of 2 from the OLT (i.e., $d_i = 2$). For each additional stage, we insert a 1:2 splitter next to the OLT and double the number of branches and attached ONUs until we reach the maximum number $N = 4096$ ONUs, which is close to the experimentally demonstrated state-of-the-art LR-PON with a total of 4352 ONUs (Talli and Townsend [2006]). For now, we assume the same failure probability for the fiber links of different stages (i.e., $p_0 = p_1 = \cdots =: p$). Figure 15.2 depicts the probability q_i that a given ONU i is connected to the OLT vs. the number of stages for different fiber link failure probability $p \in \{10^{-5}, 10^{-4}, 10^{-3}\}$. We observe that the probability of an intact optical connection to the OLT decreases for an increasing number of stages and ONUs, especially for $p = 10^{-3}$. This figure illustrates the importance of providing improved survivability in NG-PONs as their increased number of stages and ONUs result in a decreasing ONU connectivity probability q_i.

Next, we investigate the beneficial impact of interconnecting ONUs through a WMN. Figure 15.3 depicts the average number D of failure-free connections among a fixed number of $N = 1024$ ONUs vs. fiber link failure probability p, which is assumed to be the same in all eight stages of the binary tree. For simplicity, we use the random selection scheme to choose $M \leq N = 1024$ ONUs, whereby $M \in \{0, 16, 32, 64, 128, 256, 512, 1024\}$. Except for $M = N$, we observe that the average number of failure-free connections decreases for an increasing fiber link failure probability and asymptotically approaches zero for $M \leq 64$. Increasing the number of wirelessly upgraded ONUs to $M = 128$ and higher increases the number of failure-free connections. The random selection of $M = 512$, i.e., randomly equipping 50% of the

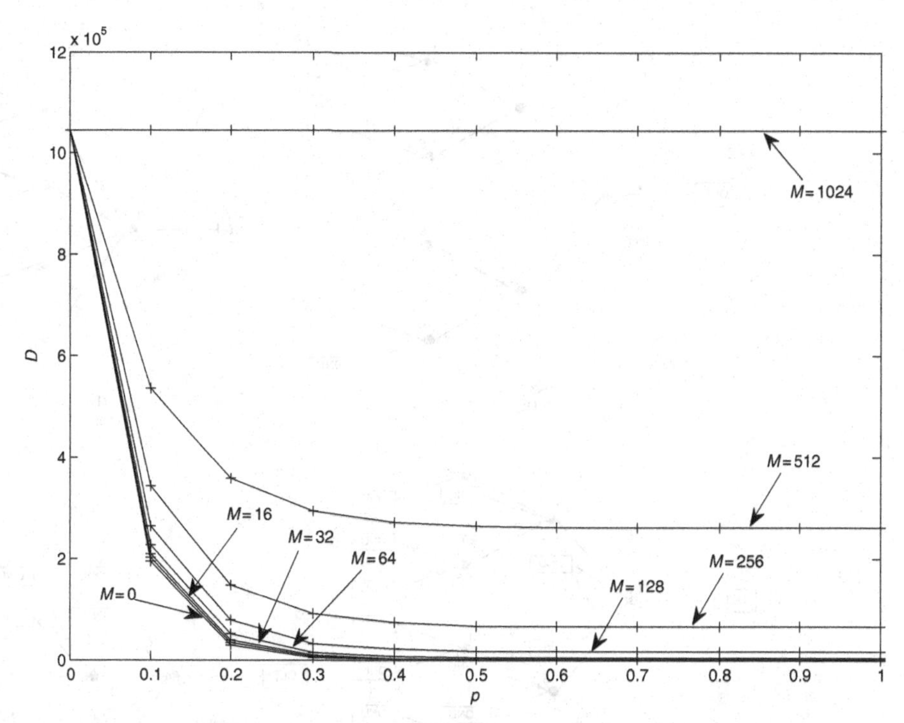

Figure 15.3 Average number D of failure-free connections among $N = 1024$ ONUs vs. fiber link failure probability p (same for all stages).

ONUs with MPPs, helps maintain roughly 25% of all connections among ONUs for a medium and high fiber link failure probability p. Note that full connectivity among all ONUs can be achieved for any value of p by equipping all $N = 1024$ ONUs with an MPP.

In the following, we examine the benefits and limitations of equipping a subset of ONUs with MPPs in greater detail by taking a number of different NG-PON topologies into account and comparing them with the above binary tree topology. Figure 15.4 shows the four different NG-PON topologies we consider for the performance evaluation of our proposed selection schemes:

- *Binary tree*: As mentioned above, the binary tree uses 1:2 splitters in all its stages except for the last one. The last stage uses 1:S splitters to connect the ONUs to the tree. Each additional stage increases the number of attached ONUs by a factor of 2. Note that in the binary tree all ONUs have the same distance to the OLT.
- *Full tree*: In the full tree, each stage deploys 1:S splitters. Similar to the binary tree, all ONUs have the same distance to the OLT. Clearly, for a given number of ONUs their distance is smaller in the full tree than in the binary tree.
- *Pyramid*: The pyramid uses only 1:S splitters, but ONUs are allowed to be located at different distances from the OLT. In the pyramid, ONUs are connected not only to the splitters of the last stage but also to intermediate splitters. Specifically, each intermediate splitter connects to $S - 2$ ONUs while the remaining two branches connect to

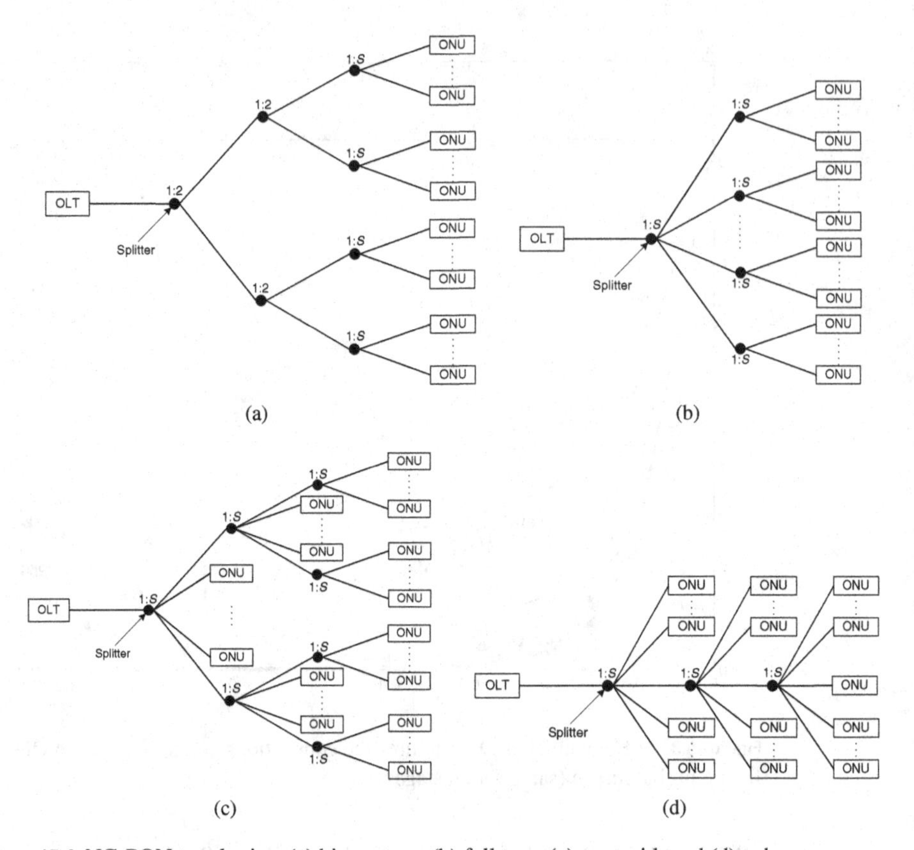

Figure 15.4 NG-PON topologies: (a) binary tree, (b) full tree, (c) pyramid, and (d) cube.

the next stage. At the final stage, each splitter connects to S ONUs. Note that, like the binary tree, each additional stage of the pyramid doubles the number of connected ONUs.

- *Cube*: Similar to the pyramid, the cube deploys only 1:S splitters and allows ONUs to have different distances to the OLT, whereby each stage increases the number of ONUs by $S - 1$ ONUs. In the cube, each intermediate splitter connects to $S - 1$ ONUs while the remaining branch connects to the next stage. The final splitter has S ONUs attached to it.

Changing the splitting ratio S of the aforementioned topologies leads to different NG-PON configurations. Figure 15.5 shows the impact of different $S \in \{16, 32, 64, 128, 256\}$ on the average number D of failure-free connections in a binary tree with $M = 64$ and $N = 1024$ vs. fiber link failure probability p (same for all stages). We observe from Fig. 15.5 that the average number of failure-free connections among ONUs increases for increasing splitting ratio S. This is due to the fact that a larger S implies that fewer stages are required to connect the 1024 ONUs to the OLT. The reduced number of required stages in addition to the fact that with an increased S there are more fiber links at the final stage, whose cuts affect only single ONUs, make the binary tree more robust against link failures, resulting in an increased number of

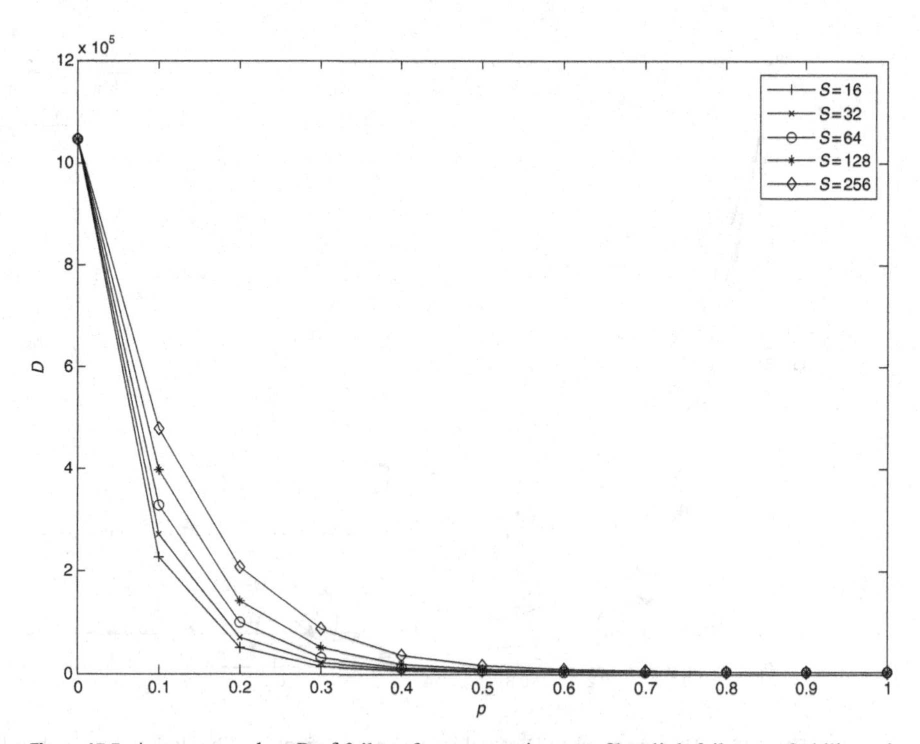

Figure 15.5 Average number D of failure-free connections vs. fiber link failure probability p in binary tree with different splitting ratio S for $M = 64$ and $N = 1024$.

failure-free connections and improved survivability. More importantly, note that the results shown in Fig. 15.5 are the same for all four different selection schemes. That is, in a binary tree we obtain the same value of D independent of the applied selection scheme. This is due to the fact that all ONUs have the same distance and are thus identical in terms of link failure probability and disconnection from the OLT. As a consequence, operators of a binary tree based NG-PON are free to choose any ONUs for a wireless upgrade in order to achieve the same level of survivability, thereby greatly simplifying network migration from NG-PON to FiWi networks. The selection of the M ONUs can be made based on the requirements of the available WMN. For instance, a network operator may select M ONUs that are close to each other in order to build a WMN with fewer or even no intermediate MPs and wireless links of shorter length. The same observations hold for our second NG-PON topology, the full tree. However, the full tree is able to achieve a significantly higher number D of failure-free connections than the binary tree, as shown in Fig. 15.6 for a fixed splitting ratio $S = 32$ and different $M \in \{0, 128, 256, 512\}$, whereby $N = 1024$.

Next, we examine the pyramid NG-PON topology where ONUs don't have the same distance to the OLT. Figure 15.7 compares the performance of our different selection schemes for a five-stage pyramid NG-PON topology with splitting ratio $S = 32$, which translates into a total number of $N = 466$ ONUs, in terms of average number D of failure-free connections vs. number M of ONU-MPPs. Furthermore, we allow each

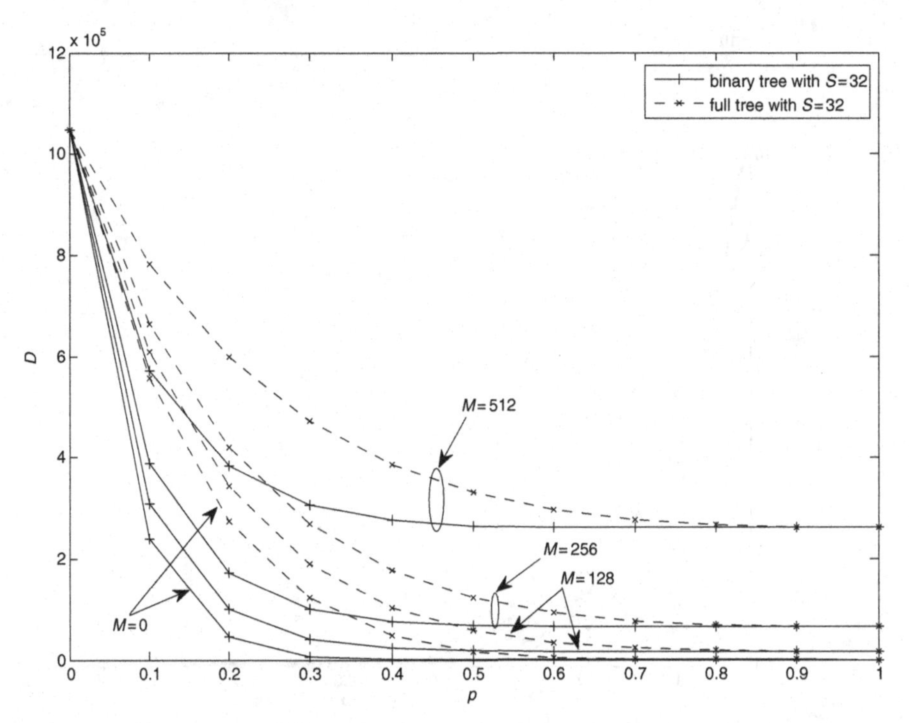

Figure 15.6 Average number D of failure-free connections vs. fiber link failure probability p in binary tree and full tree with splitting ratio $S = 32$ for different M ($N = 1024$ fixed).

stage to have a different fiber link failure probability, i.e., $p_0 \neq p_1 \neq \cdots \neq p_4$. More specifically, we consider different scenarios where the fiber link failure probability per stage is descending, ascending, or a combination thereof for an increasing distance from the OLT. In the descending scenario, the fiber link failure probability per stage decreases for an increasing distance from the OLT. Conversely, for an increasing distance from the OLT the fiber link failure probability per stage increases in the ascending scenario. In the ascending–descending scenario, the fiber link failure probability per stage increases for an increasing distance from the OLT, with the middle fiber link having the highest link failure probability, and from that link for an increasing distance from the OLT the fiber link failure probability per stage decreases. Conversely, in the descending–ascending scenario, the fiber link failure probability per stage decreases for an increasing distance from the OLT, with the middle fiber link having the lowest link failure probability, and from that link for an increasing distance from the OLT the fiber link failure probability per stage increases. In Fig. 15.7(a), the fiber link failure probability per stage decreases by a factor of 10 for an increasing distance, whereby $p_0 = 10^{-4}$, $p_1 = 10^{-5}$, $p_2 = 10^{-6}$, $p_3 = 10^{-7}$, and $p_4 = 10^{-8}$. We observe that all four selection schemes essentially show the same performance since all ONUs, independent of their distance, are equally affected by the dominating fiber link probabilities p_0 and p_1, which are by one or more orders of magnitude larger than p_2, p_3, and p_4. In contrast, a significant difference between the selection schemes can be observed in the opposite case where

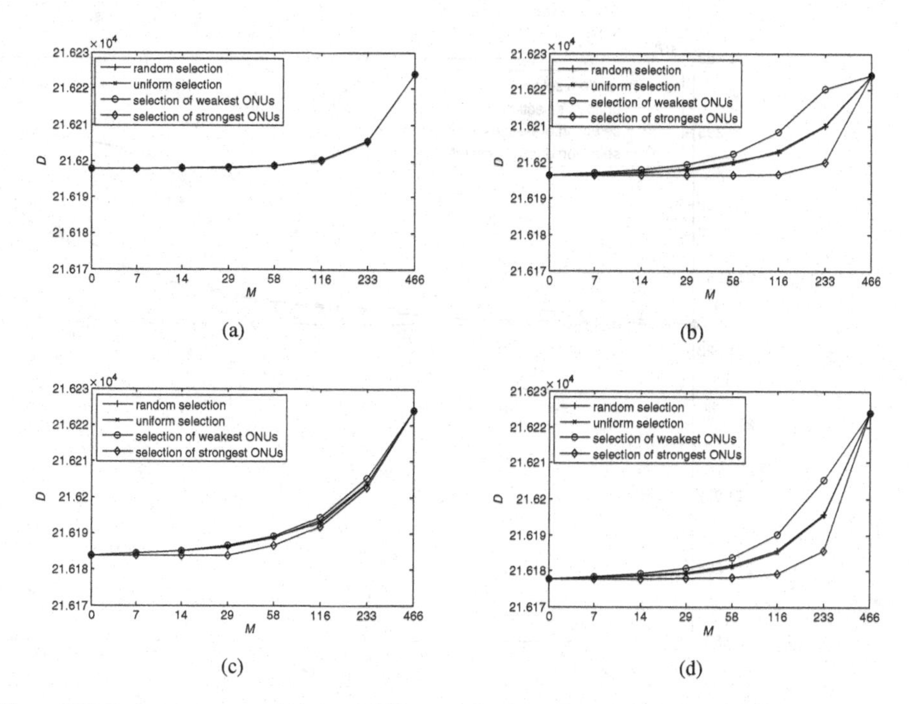

Figure 15.7 Performance comparison of different selection schemes for a five-stage pyramid NG-PON topology with splitting ratio $S = 32$ interconnecting $N = 466$ ONUs and various fiber link failure probability scenarios: (a) descending, (b) ascending, (c) ascending–descending, and (d) descending–ascending.

the fiber link failure probability per stage increases by a factor of 10 for an increasing distance, as shown in Fig. 15.7(b) for $p_0 = 10^{-8}$, $p_1 = 10^{-7}$, $p_2 = 10^{-6}$, $p_3 = 10^{-5}$, and $p_4 = 10^{-4}$. While no significant difference between the uniform and random selection schemes can be observed, Fig. 15.7(b) demonstrates that the scheme of selecting the M weakest ONUs clearly outperforms the strongest ONU selection scheme. More precisely, with the strongest ONU selection scheme, the performance gain is negligible for up to $M = 116$, i.e., even upgrading 25% of the ONUs with MPPs does not achieve any sizable survivability improvement. Conversely, with the weakest ONU selection, the number D of failure-free connections grows exponentially for increasing M, reaching almost full connectivity already for $M = 233$, i.e., 50% of the ONUs are upgraded with an MPP. In Fig. 15.7(c), the central stage is assumed to suffer from the maximum fiber link failure probability while fiber links failures become less likely towards the OLT and far distant ONUs, i.e., we have $p_0 = 10^{-8}$, $p_1 = 10^{-6}$, $p_2 = 10^{-4}$, $p_3 = 10^{-6}$, and $p_4 = 10^{-8}$. As shown in Fig. 15.7(c), we observe that such a failure scenario makes the difference between the weakest and strongest ONU selection schemes less pronounced and makes them comparable to the random and uniform selection schemes, especially for small and large values of M. In the fourth and final failure scenario under consideration, the central stage is assumed to be less failure prone than the other stages by setting $p_0 = 10^{-4}$, $p_1 = 10^{-6}$, $p_2 = 10^{-8}$, $p_3 = 10^{-6}$, and $p_4 = 10^{-4}$, as depicted in Fig. 15.7(d). Similarly to Fig. 15.7(b), the weakest ONU selection scheme is again

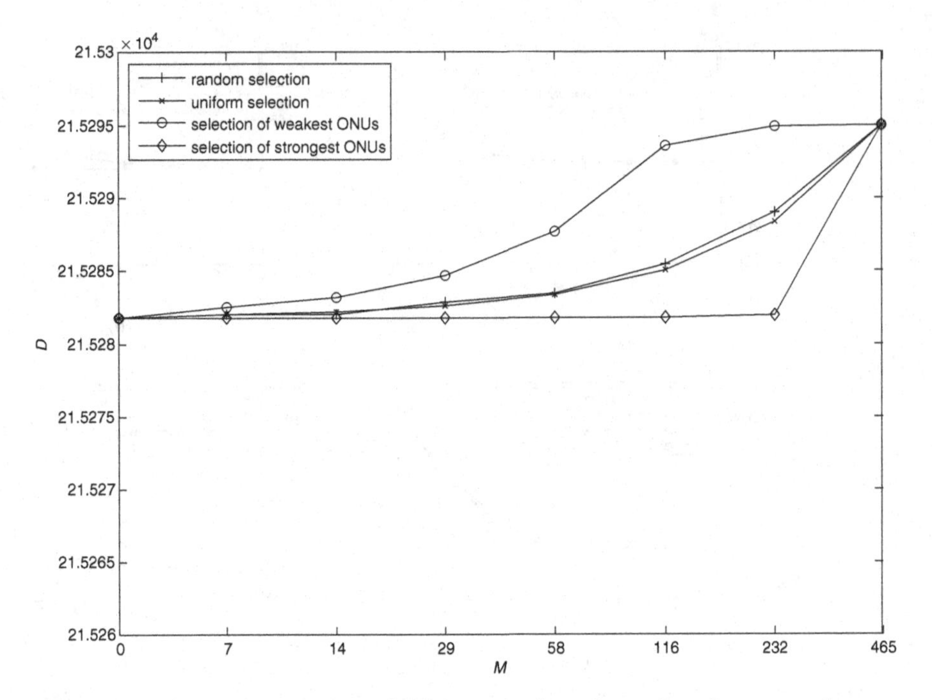

Figure 15.8 Performance comparison of different selection schemes for a five-stage cube NG-PON topology with splitting ratio $S = 117$ interconnecting $N = 465$ ONUs under the ascending fiber link failure probability scenario.

superior to the strongest ONU selection scheme. However, for all four selection schemes the number D of failure-free connections is smaller in Fig. 15.7(d) than in Fig. 15.7(b) for $M < N$.

We have also studied the impact of the four aforementioned failure scenarios (descending, ascending, ascending–descending, descending–ascending) on the performance of the cube NG-PON topology. For a fair comparison with the above pyramid topology, we have considered a five-stage cube that interconnects almost the same number of ONUs $N = 465$. Toward this end, we had to set the splitting ratio of the five-stage cube to $S = 117$. Overall, we observed the same general behavior as in the pyramid with only a few subtle differences, as highlighted in Fig. 15.8 for the ascending failure probability scenario with $p_0 = 10^{-8}$, $p_1 = 10^{-7}$, $p_2 = 10^{-6}$, $p_3 = 10^{-5}$, and $p_4 = 10^{-4}$. Figure 15.8 illustrates that the choice of the right selection scheme becomes even more important in the cube topology. While there is again no major difference between the random and uniform selection schemes, the superiority of the weakest ONU selection scheme over the strongest ONU selection scheme is more pronounced in Fig. 15.8 than in Fig. 15.7(b). The former one achieves a value of D close to the maximum for already $M = 116$, i.e., upgrading only 25% of the ONUs with an MPP, while no significant performance gain can be observed for the latter one, even by wirelessly upgrading $M = 232$, i.e., 50% of the ONUs, with an MPP. Thus, it seems that that for an NG-PON topology with a high splitting ratio, such as our considered cube with $S = 117$,

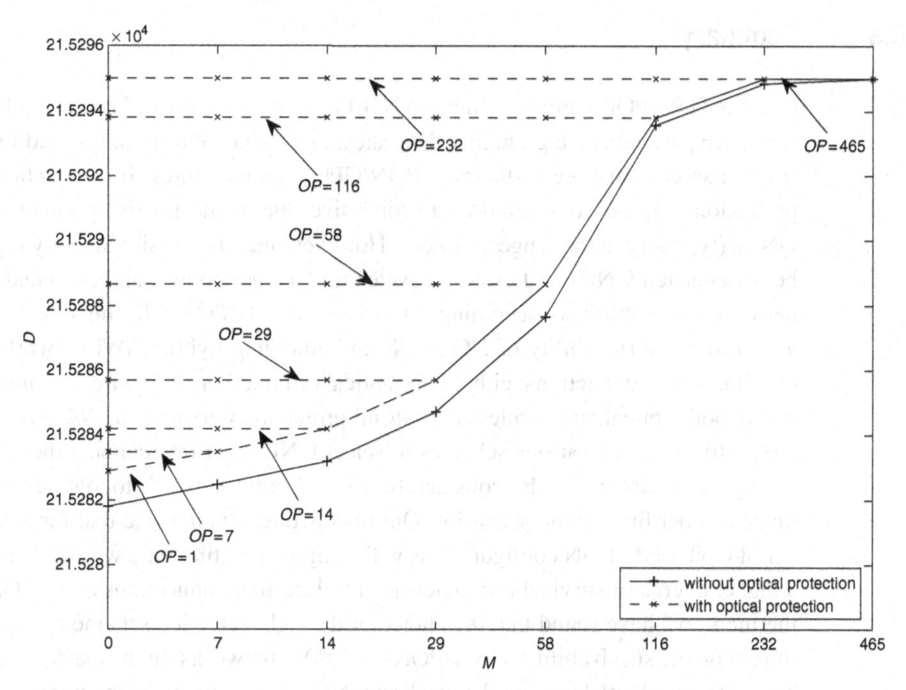

Figure 15.9 Average number D of failure-free connections vs. number M of wirelessly upgraded ONUs for a five-stage cube NG-PON topology ($S = 117$, $N = 465$) with and without optical protection.

the survivability can be improved significantly (i.e., D close to maximum) by equipping only a relatively small subset of ONUs with an MPP. In other words, an NG-PON topology with a high splitting ratio has the potential to provide a good survivability performance–cost trade-off by deploying the right selection scheme.

So far, we have assumed that there is no optical protection. Figure 15.9 depicts the beneficial impact of deploying optical protection in addition to interconnecting a subset of ONUs through a WMN for the aforementioned five-stage cube topology with $S = 117$ and $N = 465$. The figure shows the average number D of failure-free connections vs. number M of wirelessly upgraded ONUs with and without optical protection using the weakest ONU selection. More precisely, without optical protection we select the M weakest ONUs and interconnect them through a WMN, as shown above in Fig. 15.9. With optical protection, the wirelessly upgraded M ONUs are additionally optically protected by means of back-up fibers such that their optical connection to the OLT can be considered safe, i.e., $q_i = 1$. In Fig. 15.9, the number OP of optically protected ONUs is $OP \in \{1, 7, 14, 29, 58, 116, 232, 465\}$. Figure 15.9 clearly demonstrates that deploying partial optical protection in combination with using a WMN front-end helps increase D significantly. Especially for small to medium M (and OP), partial optical protection is an effective yet cost-efficient means to improve the survivability of NG-PONs and FiWi networks considerably.

15.3 Summary

Current TDM PONs are evolving into NG-PONs with the goal of achieving higher performance parameters, e.g., higher data rates, increased splitting ratios, and longer fiber reach than current state-of-the-art EPON/GPON architectures. In conventional PONs, protection may be considered cost-prohibitive due to the relatively small number of ONUs (typically in the range 16 to 64). However, due to their significantly higher number of attached ONUs, data rates, wavelength channel counts, and extended coverage, network survivability is becoming a key issue in NG-PONs. In this chapter, we have analyzed the survivability of NG-PONs and emerging hybrid FiWi networks in terms of failure-free connections, either only optical or mixed optical–wireless, among ONUs taking both optical and wireless partial/full protection into account. We have compared the performance of various schemes to select ONUs and interconnect them wirelessly through a WMN under the consideration of different network topologies and a wide range of fiber link failure scenarios. Our obtained results indicate that for a given number of ONUs NG-PON configurations with a higher splitting ratio we are able to achieve a higher degree of survivability in terms of failure-free connections among ONUs. Furthermore, we have found that the choice of the right selection scheme has a significant impact on the survivability of NG-PONs and FiWi networks. In the example of the cube NG-PON topology, by using the weakest ONU selection scheme the number of failure-free connections among ONUs is almost maximized by wirelessly upgrading only 25% of the ONUs. Finally, we have seen that partial optical protection in combination with a WMN front-end is an effective and cost-efficient means to improve the survivability of NG-PONs and FiWi networks considerably, especially for small to moderate numbers of required ONU-MPPs.

16 Hierarchical frame aggregation

We saw in Section 10.2 that significant progress has been made at the physical layer of fiber-wireless (FiWi) and in particular radio-over-fiber (RoF) networks. However, state-of-the-art radio-and-fiber (R&F) networks integrating Ethernet passive optical networks (EPONs) with a wireless local area network (WLAN)-based wireless mesh network (WMN) (see Fig. 10.4) suffer from a poor quality of video transmissions that sharply deteriorates for an increasing number of wireless hops. Therefore, a more involved investigation of the performance of integrated EPON/WLAN-based WMN networks, especially in the wireless segment, is needed.

In this chapter, we propose and investigate a FiWi network architecture that converges next-generation WLAN-based WMN and EPON networks. The considered FiWi network architecture enables existent EPON networks to be upgraded with wireless extensions in a pay-as-you-grow manner while providing backward compatibility with legacy infrastructure and protecting previous investment. Furthermore, the benefits of extending advanced frame aggregation techniques to EPON and their integrated operation across both optical and wireless segments are investigated. For more detailed information the interested reader is referred to (Ghazisaidi *et al.* [2010], Ghazisaidi and Maier [2011]).

16.1 Integration of next-generation WLAN and EPON

Figure 16.1 depicts our proposed network architecture and node structures for integrating a next-generation WLAN-based WMN with an EPON network. In this figure, an ONU represents a conventional EPON optical network unit (ONU), as described in Chapter 4. Some of the ONUs are upgraded with a mesh portal point (MPP) to interface with the WMN. The selection of ONUs that are upgraded with wireless hardware and software might be done based on various criteria, e.g., for wireless protection of the EPON. Upgrading selected ONUs with wireless equipment might be done in a pay-as-you-grow manner. In the integrated FiWi network of Fig. 16.1, the ONU MPP and mesh point (MP) play a major role.

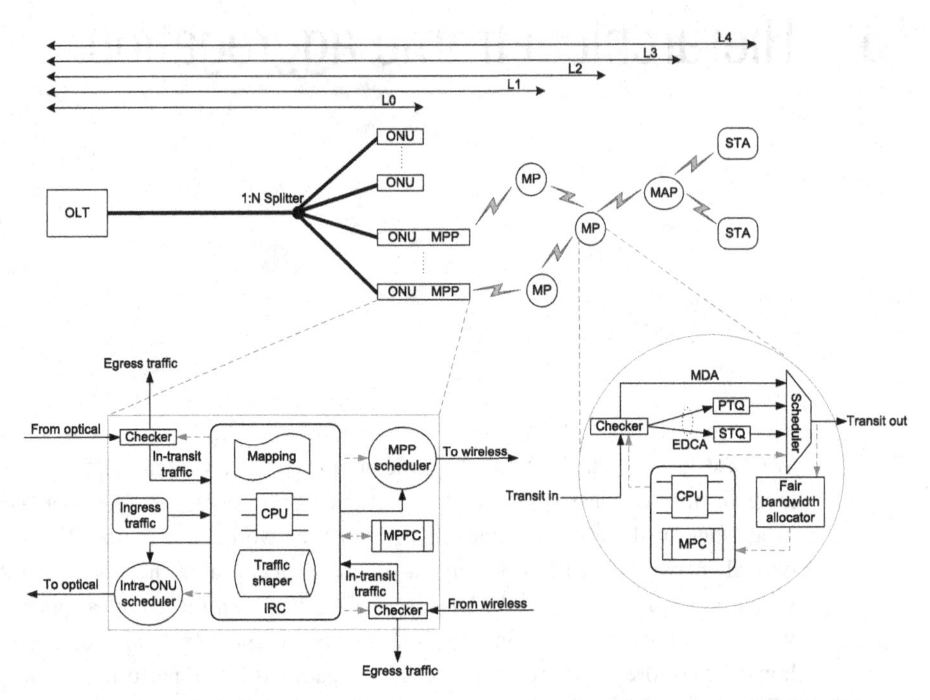

Figure 16.1 Network architecture and node structures of integrated next-generation WLAN-based WMN and EPON.

16.1.1 ONU MPP

An integrated ONU MPP node consists of the following four main modules: intra-ONU scheduler, MPP scheduler, integrated rate controller (IRC), and MPP controller (MPPC). The intra-ONU and MPP schedulers are used to transmit local and forward in-transit traffic to the EPON and WMN, respectively. The ingress traffic and egress traffic denote the traffic generated by and destined to the local ONU, respectively. The IRC plays a key role in integrating the ONU and MPP. It comprises a traffic class mapping unit, central processing unit (CPU), and traffic shaper. It is used to seamlessly integrate both technologies and jointly optimize the intra-ONU and MPP schedulers. The traffic class mapping unit translates the different EPON and wireless enhanced distributed channel access (EDCA)/mesh deterministic access (MDA) traffic classes bidirectionally. In EPON, ONUs use the priority queues defined in IEEE 802.1D to store the following seven traffic types: network control, voice, video, controlled load, excellent effort, best-effort, and background traffic (note that one queue is reserved for future purpose). While EDCA deploys four different access categories AC_VO, AC_VI, AC_BE, and AC_BK for voice, video, best-effort, and background traffic, respectively, no specific traffic classes are defined for MDA in IEEE 802.11s. The bidirectional class mapping between EPON and EDCA is done as follows: EPON network control and voice traffic is mapped to EDCA AC_VO, EPON video and controlled load traffic is mapped to EDCA AC_VI, EPON excellent and best-effort traffic is mapped to EDCA AC_BE, and finally EPON background traffic is mapped to EDCA AC_BK.

For guaranteed quality-of-service (QoS) support of EPON network control and voice traffic as well as AC_VO, ONU MPPs and MPs use MDA to reserve bandwidth in the WMN, provided there is sufficient unreserved bandwidth. Otherwise, ONU MPPs and MPs use EDCA.

The traffic shaper checks the load of each scheduler and performs traffic shaping and hierarchical aggregation/de-aggregation, as described in greater detail in Section 16.2. The CPU synchronizes the aforementioned units and controls the intra-ONU and MPP schedulers. More precisely, the role of the CPU is twofold: (*i*) synchronizing all the operational processes at different modules of the IRC, including alarm management, and (*ii*) monitoring and tuning the traffic shaper and schedulers dynamically in order to optimize QoS-aware packet delivery. Moreover, the CPU monitors the load of incoming traffic from the optical and wireless segments and synchronizes the schedulers based on traffic monitoring information. For instance, in the case of wireless network congestion, the CPU generates a congestion control message and sends it to the transmitter MP to perform load balancing. We note that the dashed lines in Fig. 16.1 represent the monitoring functions executed by the CPU. The MPPC is responsible for handling incoming and outgoing wireless traffic, beside establishing and monitoring wireless links.

16.1.2 MP

The wireless MP node structure comprises the following four main modules: MP scheduler, MP controller (MPC), fair bandwidth allocator, and CPU, which is used to monitor the aforementioned components. The MP scheduler is used to forward in-transit traffic to other wireless nodes. Incoming in-transit frames are transferred to the checker and are subsequently switched to the scheduler or transit queues based on their type of access, i.e., EDCA or MDA. The primary transit queue (PTQ) and secondary transit queue (STQ) are defined for prioritized transmission of EDCA packets. More specifically, the PTQ is dedicated to the two high-level priority access categories (i.e., AC_VO and AC_VI) and the STQ is used by the others (i.e., AC_BE and AC_BK). MDA in-transit frames are processed by the scheduler without queueing and pass the MP without de-aggregation. In contrast, EDCA frames can be de-aggregated and re-ordered based on the priority of the transit queues. The fair bandwidth allocator module monitors the reserved MDA bandwidth and processes control congestion packets. The MP, which receives many MDA opportunity (MDAOP) requests from its neighborhood MPs, sends a *fair bandwidth alarm* message. As a result, the MP, which receives the alarm, should find another MDA path to the destination or applies AC_VO for transmitting high-priority traffic in the EDCA mode. Note that the above-mentioned transit queues and fair bandwidth allocator module are inspired by IEEE 802.17 resilient packet ring (RPR) for its simplicity and cost-efficiency (Davik *et al.* [2004]). The CPU synchronizes the aforementioned units and controls the MP scheduler and checker. The MPC is responsible for handling incoming and outgoing traffic, beside avoiding local congestion. The mesh access point (MAP) node structure is similar to the MP node structure, but additionally provides service to wireless stations (STAs).

We note that our considered FiWi network is not restricted to any specific scheduling algorithm for the ONU MPP and MP. ONU MPPs and MPs perform channel access, bandwidth reservation, relaying, and routing according to the extensible routing protocol framework specified in IEEE 802.11s. They periodically monitor the reserved MDA bandwidth and current traffic loads in the WMN for path computation and load balancing. In the WMN, ONU MPPs and MPs apply the hybrid wireless mesh protocol (HWMP), which aims at combining ad-hoc on-demand distance vector (AODV) and tree-based pro-active routing algorithms.

16.2 Hierarchical frame aggregation techniques

We apply an aggregate MAC service data unit (A-MSDU) in the optical segment since for an error-free medium such as optical fiber A-MSDU achieves a higher throughput than an aggregate MAC protocol data unit (A-MPDU) (Lin and Wong [2006]), as explained in Section 6.3.1. The A-MPDU is used in the error-prone WMN. We note that the proposed frame aggregation techniques apply only to EDCA frames. As shown in Fig. 16.1, our proposed hierarchical frame aggregation techniques involve the following five different aggregation layers L0–L4:

- **L0:** This layer applies A-MSDU for traffic between the optical line terminal (OLT) and conventional ONUs as well as for ingress and egress ONU MPP traffic going to and coming from the OLT.
- **L1:** This layer performs hop-by-hop aggregation for traffic between the OLT and first-hop MPs. A-MSDU is used for traffic between the OLT and ONU MPPs, whereby an ONU MPP additionally applies A-MPDU for the first wireless hop. More specifically, in the downstream direction an ONU MPP combines A-MSDUs destined to the same first-hop MP and encapsulates them into A-MPDUs (i.e., joint two-level aggregation). In the upstream direction, an ONU MPP first de-aggregates incoming A-MPDUs arriving from one or more first-hop MPs and then re-orders and aggregates the individual MSDUs into A-MSDUs based on their traffic identifier (TID).
- **L2:** By default, all MPs are involved in relaying in-transit frames, but only a subset of MPs are allowed to perform de-aggregation, re-ordering, aggregation of in-transit frames. We refer to these special MPs as *virtual root MPs (VRMPs)*. In general, any MP between ONU MPPs and MAPs can be a VRMP. In the downstream direction, a VRMP carries out two-level de-aggregation, re-ordering, and single-level A-MPDU aggregation of frames destined to the same MAP. Note that a VRMP utilizes only A-MPDU for aggregation due to its superiority on error-prone wireless links. In the upstream direction, a VRMP first de-aggregates A-MPDUs coming from one or more MAPs, re-orders the individual MSDUs based on their destination address and TID, and then performs two-level aggregation (i.e., joint A-MSDU and A-MPDU) of in-transit frames going to the OLT. MPs located between a VRMP and an ONU-MPP relay the two-level frame aggregates and are not allowed to de-aggregate them. At the

ONU-MPP, two-level aggregates coming from a VRMP are translated into their constituent A-MSDUs, which are subsequently forwarded to the OLT. The selection of VRMPs has a major impact on the network performance, as we will see shortly in Section 16.4.

- **L3:** This layer combines L2 aggregation of frames exchanged between the OLT and a given VRMP with hop-by-hop A-MPDU on the wireless link(s) between each MAP and the given VRMP.
- **L4:** This layer is designed for end-to-end aggregation of traffic between the OLT and STAs. A-MPDU is used on the wireless link between each STA and its associated MAP, which in turn performs L3 aggregation for communication to and from the OLT.

16.3 Capacity of wireless mesh front-end

In the WMN, we let N nodes, i.e., MPPs, MPs, or STAs, be uniformly distributed. In an error-free single-channel WMN, a pair of nodes are connected if and only if the distance between them is less than their maximum radio transmission range to satisfy the constraints of EDCA or MDA. For instance, the maximum distance between two connected nodes is less than 2.7 km in IEEE 802.11n with a slot size of 9 μs and short interframe space (SIFS) of 16 μs. In our network model, the simultaneous data transmission on two different edges $e = \{i, j\}$ and $\tilde{e} = \{k, l\}$ is possible if and only if e and $\tilde{e}$ are sufficiently far apart from each other, i.e., none of the nodes i and j is adjacent to any of the nodes k and l. We note that edges are undirected, i.e., we write $\{i, j\}$ rather than (i, j).

For simplicity, we consider shortest path routing in terms of number of hops. Let λ_{ij} be the traffic rate from WMN node i to j scaled such that $\lambda_{ij} = 1$ corresponds to the maximal capacity of the edges. For each edge $e = \{i, j\}$, we determine the total traffic in both directions $t_{\{i,j\}}$ by summing up all λ_{kl} for nodes k and l that use edge e. Let ξ denote the set of all edges. We call $\zeta \subseteq \xi$ *admissible*, if any two different edges in ζ are far apart. Note that a set of edges is admissible if and only if all edges of the set can be used simultaneously for data transmission. We say that $\zeta \subseteq \xi$ is *maximally admissible*, if ζ is admissible and $\zeta \subseteq \tilde{\zeta} \subseteq \xi$, where the admissibility of $\tilde{\zeta}$ implies $\tilde{\zeta} = \zeta$.

In a stable system, there exists a maximally admissible set ζ such that only edges of that set are used for each data transmission. We note that ζ may not be unique. If α_ζ denotes the long-time probability in which ζ is used, then the stability condition is $\sum_\zeta \alpha_\zeta \leq 1$. The network stability means that each edge $e \in \xi$ can cope with the traffic t_e that has to go through e. Therefore, a necessary condition for network stability is that for each maximally admissible ζ, there exists $\alpha_\zeta \geq 0$ such that

$$\begin{cases} \sum_\zeta \alpha_\zeta = 1 \\ \sum_{e \in \zeta} \alpha_\zeta \geq t_e \end{cases}. \tag{16.1}$$

Linear programming can be used to solve the stability limits of Eq. (16.1) and thus obtain the capacity of WMN.

16.4 Numerical and experimental results

In our first set of simulations, we consider uniform unicast traffic between OLT, ONU MPPs, and STAs. We assume that 48 STAs are connected to the OLT through 80 MPs and 32 ONU MPPs, whereby the distance between ONU MPPs and OLT is set to 20 km. The STAs are located at a range of 2 km of the associated MPs, while the distance between a connected pair of MPs and MPPs is set to 2 km. The optical line rate is set to 1 Gb/s. We consider Poisson data traffic with different packet sizes equal to 40, 552, and 1500 bytes according to a distribution of 50%, 30%, and 20%, respectively, transmitted with an additional 20-byte transmission control protocol (TCP) header and 20-byte Internet protocol (IP) header. In addition, we use the voice codec standard ITU-T G.711 with a constant bit rate (CBR) source rate of 64 kb/s, where each packet contains 12, 8, and 20 bytes of real-time transport protocol (RTP), user datagram protocol (UDP), and IP headers, respectively. We deploy MPEG-4 video codec to encode 600-byte packets at a data rate of 768 kb/s, which generates UDP CBR traffic, including 8 bytes and 20 bytes of UDP and IP headers, respectively. In our simulations, the two voice and video codecs are used simultaneously, each encoding 50% of generated traffic, while Poisson traffic uses 20% of the network capacity. In EPON, we use the limited-service interleaved polling with adaptive cycle time (IPACT) with a maximum grant size of G_{max}=15 kbytes as the dynamic bandwidth allocation (DBA) algorithm (Kramer *et al.* [2002a]). In WLAN, the bit error rate (BER) of the wireless channel is set to 10^{-5}.

In Fig. 16.2, the analysis results show the upper-bound of network throughput (i.e., capacity) verified by simulation. The simulations yield smaller throughput results than the analysis due to the realistic wireless channel conditions and applying a request to send/clear to send (RTS/CTS) packets in WMN as well as REPORT/GATE packets in

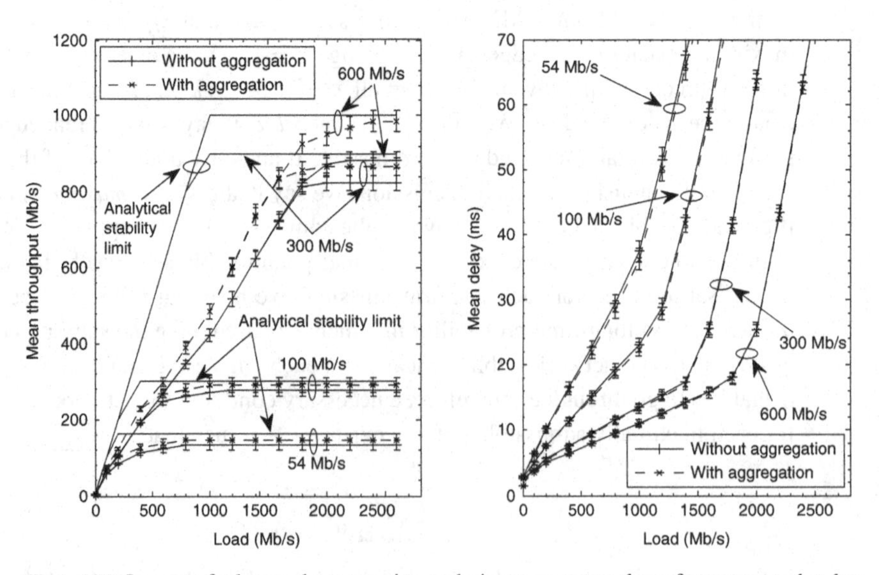

Figure 16.2 Impact of advanced aggregation techniques on network performance under data traffic.

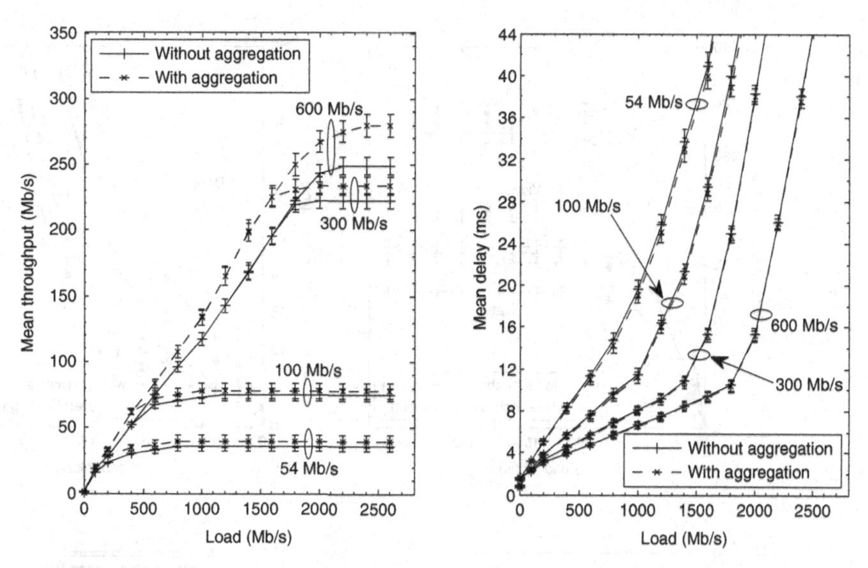

Figure 16.3 Impact of advanced aggregation techniques on network performance under triple-play traffic.

EPON. Furthermore, Eq. (16.1) shows the stability limits of WMN for a flow of packets, while the limited-service granting approach is used in our simulations. As result, we observe that the simulation results of the proposed aggregation techniques (with smaller overhead) are closer to the analysis results than without aggregation, including larger overhead.

Figure 16.3 shows the impact of our proposed aggregation techniques on the performance of the FiWi network for various WMN data rates under triple-play traffic (i.e., voice, video, and data). In this figure, mean throughput and mean delay are averaged over all three traffic types.

In the remaining simulations, we assume 32 ONUs to be located at 20 km from the OLT, whereby 16 ONUs are equipped with an MPP. Furthermore, we assume that two STAs are located at 30 m from their associated MAP and the distance between a pair of adjacent MAP, MP, and ONU MPP equals 1 km. Each STA connects to an ONU MPP via an MAP and two intermediate MPs. In L3 of Fig. 16.1, each ONU MPP connects to two MPs and each MP is in the transmission range of two ONU MPPs. In total, there are 16 ONUs, 16 ONU MPPs, 32 MPs, 16 MAPs, and 32 STAs. The EPON and WMN data rates are set to 1 Gb/s and 100 Mb/s, respectively. We consider the uniform unicast triple-play traffic from above, where a given node (i.e., OLT, ONU, ONU MPP, or STA) sends a generated packet to any other node with equal probability $1/(N-1)$ and N denotes the number of nodes.

Figure 16.4 illustrates the beneficial impact of our hierarchical frame aggregation on the throughput-delay performance of integrated EPON-WLAN based FiWi networks for voice, data, and especially video traffic. The 95% confidence intervals exhibit an error of less than 10% from the mean values. We examine the performance of two different VRMP selection schemes. In Fig. 16.4(a), we apply the so-called *MAP neighborhood*

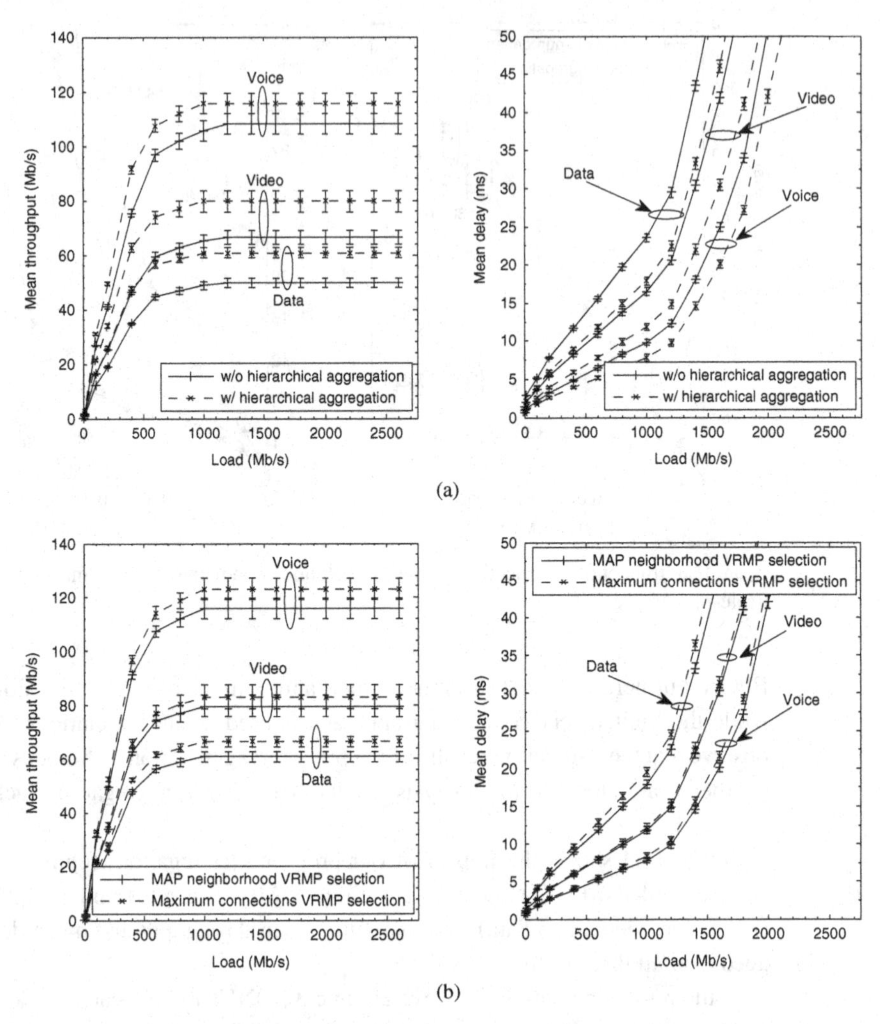

Figure 16.4 Impact of hierarchical frame aggregation on integrated EPON-WLAN network performance under triple-play (voice, video, and data) traffic: (a) using the MAP neighborhood VRMP selection scheme and (b) comparing two different VRMP selection schemes.

VRMP selection scheme, which selects the MPs next to the MAPs as VRMPs, and show its superior performance to an EPON-WLAN network without using our hierarchical aggregation techniques. In Fig. 16.4(b), we compare the MAP neighborhood VRMP selection scheme with another scheme, which we call the *maximum connections VRMP selection* scheme. The latter one selects the MPs with the maximum number of wireless connections, which happen to be located next to the ONU MPPs in our scenario. Figure 16.4(b) shows that the first scheme is able to achieve a smaller mean delay, whereas the second one may be used to further increase the mean throughput. Thus, the throughput and delay of the network can be improved by selecting VRMPs at different locations.

Next, we experimentally investigate frame aggregation in both the optical and wireless segments. Our FiWi testbed consists of a Sun Telecom GE8200 EPON with four

Table 16.1. Experimental results: Impact of hierarchical frame aggregation techniques on FiWi network performance for 60 second VoIP connection.

Optical	Wireless	Mean jitter	Mean packet loss	MOS
w/o Aggregation	w/o Aggregation	4.7	0.2	3.5
w/o Aggregation	w/ A-MSDU	3.7	0.1	3.9
w/o Aggregation	w/ A-MPDU	2.5	0.1	4.3
w/o Aggregation	w/ Joint Two-level	2.4	0.09	4.4
w/ A-MSDU	w/ Joint Two-level	2.1	0.09	4.6

ONUs located at 20 km from the OLT. One ONU interfaces with a ZyXEL NWA570N next-generation WLAN access point to realize an ONU MPP. We measured the performance in terms of mean jitter (given in ms), mean packet loss (given in %), and *mean opinion score (MOS)* (MOS [2009]) for a 60 second voice over Internet protocol (VoIP) connection between the OLT and a laptop computer (STA) at 30 m from the ONU MPP using MyConnection PC.[1] Table 16.1 demonstrates that applying frame aggregation in each segment of the network improves the performance. Interestingly, we observe that A-MSDU decreases packet loss and increases throughput due to the small VoIP packet sizes over a wireless channel with small packet error rate (PER). Furthermore, the table shows that deploying A-MSDU in EPON helps decrease mean jitter and increase MOS considerably.

16.5 Summary

We proposed and examined advanced hierarchical frame aggregation techniques to improve the throughput-delay performance of an integrated WLAN-mesh/EPON based FiWi network for voice, video, and data traffic. In our hierarchical frame aggregation techniques, we applied A-MSDU in the error-free optical segment and both two-level A-MSDU/A-MPDU and A-MPDU in the error-prone WMN. The use of adaptive frame aggregation techniques according to instant traffic loads and wireless channel conditions (such as BER) would be a promising approach to further improve the performance of our proposed FiWi network. We evaluated the capacity of WMNs through probabilistic analysis and verifying simulations and investigated the performance of a FiWi network consisting of an EPON and WLAN-based WMN. The obtained results show that the proposed aggregation techniques improve the throughput-delay performance of the FiWi network under data and triple-play traffic. Our simulation results showed that the selection of VRMPs has a major impact on the network performance. Moreover, our experimental results showed that deploying hierarchical frame aggregation techniques improves the mean jitter, mean packet loss, and MOS of FiWi networks, while extending A-MSDU to EPON further decreases mean jitter and increases the MOS.

[1] MyConnection PC, 2010; http://www.myconnectionpc.com

17 Routing and QoS continuity

The wireless mesh front-end of fiber-wireless (FiWi) access networks provides multiple paths to route traffic coming from and going to the optical backhaul. In this chapter, we review a variety of recently proposed routing algorithms that aim at optimizing the network performance in terms of delay, throughput, packet loss, load balancing, and other important metrics such as path availability and power consumption. The considered routing algorithms cover either only the wireless front-end or both the wireless and optical domains of FiWi access networks. Furthermore, this chapter elaborates on various techniques to provide service differentiation and end-to-end guaranteed quality-of-service (QoS) and enable QoS continuity across the optical–wireless interface of FiWi broadband access networks.

17.1 Wireless routing algorithms

In this section, we describe various recently proposed routing algorithms for the wireless front-end of FiWi access networks. All of the discussed wireless routing algorithms aim at finding the optimal path through a wireless mesh front-end by meeting one or more objectives.

17.1.1 DARA

A pro-active routing algorithm, referred to as the *delay-aware routing algorithm (DARA)*, which minimizes the average packet delay between a router and any wireless mesh gateway was presented in (Sarkar *et al.* [2007b, 2008]). DARA is a link-state routing algorithm, where each wireless mesh router and gateway periodically advertises their link conditions (i.e., traffic intensity and link capacity with time stamp) in link state advertisement (LSA) messages. Upon reception of an LSA message, each router/gateway predicts the traffic intensity, which is used to predict the state of the corresponding link until the next LSA message arrives. To make sure that the resultant link state prediction (LSP) is accurate, the authors proposed to use *weighted moving average (WMA)* to estimate the traffic intensity of a given link between the current LSA message and the next LSA message. Based on the LSP information, each wireless link is assigned a weight according to its predicted transmission, synchronization, and queuing delay. More precisely, links with higher predicted delays are given higher weights and vice versa. Subsequently, for each pair of router and gateway a single (or alternatively $K > 1$ for load balancing) minimum-weight path(s) is (are) computed. In addition, DARA performs admission control, where a packet is admitted into the wireless mesh only if the predicted delay along the computed path is below a predetermined threshold. Otherwise, the packet is rejected.

The presented simulation results indicate that DARA outperforms other routing algorithms such as minimum hop routing, shortest path routing, and predictive throughput routing in terms of delay. Furthermore, DARA is able to improve load balancing and alleviate congestion compared to minimum hop and shortest path routing algorithms. It also improves the average hop count compared with the predictive throughput routing algorithm.

17.1.2 DDRA

The aforementioned DARA routing algorithm does not allow for any sort of QoS differentiation. To do so, the so-called *delay-differentiated routing algorithm (DDRA)* was introduced in (Chen *et al.* [2010]). The objective of DDRA is to decrease the delay of upstream transmissions to a server located at the optical line terminal (OLT) of a passive optical network (PON). Toward this end, the authors define the delay-to-server (DTS) as the time from a wireless mesh network client to the server. Traffic is divided into DTS-sensitive and DTS-insensitive traffic. Given that a significant proportion of the path delay generally occurs at the wireless bottleneck links between gateways and their adjacent wireless mesh routers, the delay performance can be improved by attaching external buffers to these routers. The external buffers can store DTS-insensitive data packets when links are congested and forward them to the server when the links to the gateways become less congested. In doing so, under high traffic loads DDRA decreases the queuing delay of DTS-sensitive traffic by temporarily storing DTS-insensitive data packets in external buffers. A given ingress wireless router computes a path to a gateway, if the traffic is DTS-sensitive. In the case of DTS-insensitive traffic, a path from the ingress router to one of the external buffers is computed. Specifically, the path to a gateway will be selected based on its computed delay. Whereas the path to an external buffer is chosen by taking both the delay and size of the external buffer into account.

The performances of DDRA and DARA were compared by means of simulation. The results reported in (Chen *et al.* [2010]) show that under light loads DDRA and DARA achieve roughly the same delay performance for DTS-sensitive traffic, though DTS-insensitive traffic suffers from a slightly higher delay. Under increasing traffic loads, DDRA outperforms DARA in terms of delay and can support nearly 1.2 times the maximum load that DARA can handle, at the expense of additional hardware requirements at some of the wireless mesh routers.

17.1.3 CaDAR

The capacity of wireless mesh networks is limited. This is particularly true for single-radio wireless mesh networks, where each mesh node is equipped with only one radio, whose capacity needs to be distributed among all the node's outgoing links by means of time division multiple access (TDMA). By assigning time slots of different duration to a node's outgoing links according to given traffic demands, each of these links has a different capacity. In (Reaz *et al.* [2007, 2008a]), a new routing algorithm, called *capacity and delay aware routing (CaDAR)*, was proposed with the objective of distributing the radio capacity of a single-radio wireless mesh network node optimally among its outgoing links such that the average delay is minimized. Specifically, in CaDAR, each node informs its neighboring nodes about the current traffic load of all its outgoing links via periodic LSA messages. Based on the advertised link loads, the CaDAR algorithm determines the fraction of a node's radio capacity that should be assigned to each of its outgoing links. Subsequently, each link is assigned its delay as weight and CaDAR computes the shortest delay path for each pair of wireless mesh router and gateway. Similar to DARA, CaDAR may deploy admission control and WMA to estimate the traffic load of each link. It was shown for a 25-node single-radio IEEE 802.11g wireless local area network (WLAN) based mesh network that CaDAR can support nearly three times the maximum load DARA can handle while maintaining low delays.

To increase the capacity of the wireless front-end and better match it to the optical backhaul, one promising approach is to deploy *multiple radios* at nodes that carry higher traffic volumes. In (Reaz *et al.* [2008b]), the problem of optimum placement of a second radio at selected wireless routers/gateways was formulated as an integer linear program (ILP) with the objective of maximizing the throughput-delay ratio of the wireless front-end. By means of simulation it was shown that the maximum throughput performance of CaDAR can be almost doubled if only the gateways deploy an additional second radio while leaving all mesh routers with a single radio. Equipping some of the wireless mesh routers with an additional radio does not lead to an increased maximum throughput due to the fact that the bottleneck links remain the same; however, it helps reduce the average delay of the network to some extent. In fact, the throughput-delay ratio of the wireless front-end increases for a growing number of multiple-radio mesh routers.

In a wireless front-end with multiple-radio mesh nodes, it is important to assign the wireless channels efficiently such that contention is reduced. A channel assignment scheme that balances traffic loads among wireless channels to minimize contention was investigated in (Reaz *et al.* [2008c]). The proposed channel assignment scheme performs load balancing of traffic flows across different channels in order to maintain a similar level of contention on each channel. Using CaDAR for routing, it was shown by means of simulation that the proposed flow-aware channel assignment scheme effectively balances the load across different wireless channels, resulting in a reduced contention throughout the wireless front-end.

17.1.4 RADAR

In FiWi access networks, packet loss may occur due to various failures, e.g., gateway or OLT failure. To cope with packet loss, the so-called *risk-and-delay aware routing (RADAR)* algorithm was studied in (Sarkar *et al.* [2007b]). RADAR is an extension of the above-mentioned DARA routing algorithm. Similar to DARA, each wireless router/gateway advertises the wireless link states in periodic LSA messages and assigns weights to wireless links according to their delays for minimum delay path computation. Unlike DARA, each mesh router maintains a *risk list* to keep track of failures. If a failure occurs, the risk list is updated and subsequent packets are rerouted. Specifically, a risk list contains the following six fields:

- Path number
- Primary gateway group (PGG)
- Secondary gateway group (SGG)
- Tertiary gateway group (TGG)
- Path status
- Path delay

whereby the path status can be either live or stale. The primary gateway for a given mesh router is the gateway with the minimum delay path. The PGG contains paths to the primary gateway and other gateways connected to the same optical network unit (ONU) of a back-end PON. The SGG contains paths to gateways that are connected to different ONUs but the same OLT as the PGG. The TGG contains paths to gateways that are connected to a different OLT.

In the event of one or more FiWi network failures, the risk list of affected mesh routers is updated accordingly. More precisely, if the primary gateway fails, then all the paths to that gateway become stale and packets destined for the primary gateway are rerouted through live PGG paths. If the ONU of the PGG paths fails, all PGG paths become stale and packets are rerouted along live SGG paths. If the common OLT of the SGG paths fails, all SGG paths become stale and packets are rerouted along live TGG paths.

RADAR provides protection against multiple wireless front-end and/or optical back-end network failures. It minimizes the average packet delay in the wireless front-end and reduces the packet loss of FiWi access networks significantly. It was shown by means of simulation that in terms of both average delay and packet loss RADAR clearly outperforms minimum hop, shortest path, and predictive throughput routing algorithms for a wide range of failure scenarios such as gateway, ONU, and OLT failures.

17.2 Integrated routing algorithms

This section reports on recent progress on integrated routing algorithms for FiWi access networks. Integrated routing algorithms compute paths across the optical–wireless interface by taking both the wireless and optical domains into account.

17.2.1 CaDAR with optical delay awareness

The CaDAR routing algorithm, originally intended only for the wireless front-end as explained above in Section 17.1.3, can be extended to perform integrated path computation across both the wireless and optical domains of FiWi access networks.

In (Reaz *et al.* [2009]), CaDAR was used in conjunction with interleaved polling with adaptive cycle time (IPACT) with constant credit service as dynamic bandwidth allocation (DBA) algorithm in a backhaul Ethernet PON (EPON). Recall from Section 4.3.1 that IPACT arbitrates the communication between OLT and ONUs. For downstream data transmissions to a wireless destination node, the OLT selects a gateway such that data packets travel on the minimum delay path across the EPON and wireless mesh front-end. To select the correct gateway, all gateways periodically send LSA messages to the OLT and the OLT computes the minimum delay path across the EPON and wireless mesh front-end. To calculate the minimum delay path for upstream data transmissions from wireless mesh clients to the OLT, a gateway takes the current load of its collocated ONU into account and estimates the upstream delay on the EPON. Subsequently, the gateway disseminates the estimated delay throughout the wireless mesh front-end via periodic LSA messages. A given wireless mesh source node uses this information for computation of the minimum delay path to the OLT.

The presented simulation results investigated the average packet delay for CaDAR under the assumption of equally and unequally loaded ONUs. It was shown that, when the load of a particular ONU is increased, more traffic is diverted toward gateways collocated with other ONUs. When all ONUs have equal load, traffic from wireless mesh source nodes tends to go toward the closest gateway using fewer wireless hops, which translates into shorter delays.

17.2.2 Availability-aware routing

Apart from capacity, one of the major differences between the optical and wireless parts of FiWi access networks is the fact that optical wired links are much more reliable than their wireless counterparts, whose transmission characteristics depend on a number of different parameters. This major difference between optical fiber and wireless links should also be taken into account in the design of integrated routing algorithms for FiWi access networks.

An *availability-aware routing* algorithm for PON/wireless mesh based FiWi networks was examined in (Shao *et al.* [2010]). Availability is the probability that a connection will be found in the operating state at a random time in the future. It is given by

$$Availability = \frac{MTTF}{MTTF + MTTR}, \tag{17.1}$$

where *MTTF* denotes the mean time to failure and *MTTR* denotes the mean time to repair. Availability is in general affected by many factors and is an important parameter in the service level agreement (SLA) between network operators and customers. The availability of a path can be calculated as the product of the availabilities of the individual links, whereby the most available path is defined as the path with the highest availability from the source node to the destination node of a given connection request. The performance of the availability-aware routing algorithm was investigated for different link availabilities and compared with shortest path routing under uniform traffic. It was shown by means of simulation that the availability-aware routing algorithm is able to achieve a higher average availability for the selected paths than the shortest path routing, especially at low traffic loads. However, availability-aware routing is less bandwidth-efficient than shortest path routing due to an increased mean number of hops.

A link availability model for FiWi access networks was developed in (Kiese *et al.* [2009]) and its performance was analyzed for various routing algorithms, including DARA, CaDAR, availability-aware routing, and multi path routing. The obtained results show that shortest path routing using link unavailability as the link cost metric significantly improves the availability of the selected paths. Moreover, it was shown that multipath routing helps further improve the availability compared with single-path routing.

17.2.3 Multipath routing

Multipath routing and its implications have been studied in greater detail in (Wang *et al.* [2010]). With multipath routing, data packets can be sent upstream to the OLT along multiple paths in the wireless front-end, resulting in improved load balancing and fault tolerance as well as higher network throughput. However, these data packets may arrive at the OLT out of order due to the fact that each path in general has a different delay. As a consequence, the out-of-order packets deteriorate the performance of the transmission control protocol (TCP). In PON-based FiWi access networks, the OLT serves as a convergence node of all upstream data flows and may as such resequence arriving packets to some extent, thereby mitigating the detrimental impact of out-of-order packet delivery and increasing TCP throughput.

In (Wang *et al.* [2010]), the authors proposed an optimal flow assignment and fast packet resequencing algorithm in order to reduce the out-of-order probability of upstream packets injected by the OLT to the Internet. In this approach, packets belonging to the same flow are sent to the OLT along multiple paths. The delay along different paths may be different and may vary over time. The OLT maintains two first-in-first-out (FIFO) priority queues with preemptive priority scheduling. Out-of-order packets are stored in the high-priority queue, while the remaining packets are put in the low-priority queue. Flows are optimally assigned among the different paths with the objective of minimizing the out-of-order probability beyond the OLT subject to given packet delay requirements.

17.3 Energy-aware routing

FiWi networks and access networks in general suffer from a major shortcoming. In today's Internet, the total energy consumption is dominated by access networks and as access rates of tens of Mb/s become commonplace, it will be necessary to improve their energy efficiency in order to avoid a significantly increased greenhouse footprint of the Internet (Tucker *et al.* [2009]). "Green" energy-efficient FiWi access networks can be realized by putting selected nodes, e.g., ONUs and wireless mesh routers, in a low-power (sleep) state in a coordinated manner during low traffic load hours in order to save energy (Mukherjee and Chowdhury [2009]).

A couple of interesting techniques to design green EPON/WLAN-mesh based FiWi access networks were studied in (Chowdhury *et al.* [2009b, 2010]). The authors proposed a centralized sleeping mechanism that allows the OLT to put low-load ONUs to sleep. Specifically, the OLT maintains two watermarks for the traffic load at each ONU, one low and one high watermark. The OLT observes the traffic load at each ONU by monitoring the length of the corresponding queues. If a given ONU operates under the low watermark, the OLT puts it to sleep and the wireless front-end is used to reroute traffic to other active ONUs. The OLT wakes a sleeping ONU up when the traffic load of the active ONUs crosses the high watermark in order to cope with the increased traffic load. The authors formulated this problem as a mixed integer linear program (MILP) to determine the minimum number of ONUs that need to be kept active to support a given traffic load, while all remaining ONUs are put in the sleep mode, thereby minimizing the power consumption of the FiWi access network.

The model can be solved for small FiWi network instances. However, for FiWi access networks with a larger number of nodes and higher traffic loads, heuristics are needed to solve the problem. Toward this end, the authors investigated an *energy-aware routing* algorithm. The objective of the proposed energy-aware routing algorithm is to use already existent paths and put the other FiWi network segments to sleep. To do so, the residual capacity of each link is assigned as its weight and LSA messages are periodically sent to inform all other network nodes about the current residual capacity of each link. Based on this information, data packets are routed along the lowest residual capacity path between each pair of source and destination nodes. Note, however, that this approach increases the average path length and path delay since the algorithm in general provides routes that are longer than the corresponding shortest paths. To mitigate this problem, the authors proposed to use a so-called *hop offset*. Given a path of n hops, for each hop a hop offset $m \geq 1$ is added to the path cost, i.e., $n \times m$ is added to the original cost of the path. The presented simulation results indicate that a significant number of ONUs of an EPON/WLAN-based mesh FiWi access network can be put to sleep without compromising its delay performance.

17.4 QoS continuity

A plethora of enabling optical and wireless network technologies and QoS provisioning techniques exist for FiWi access networks. In this section, we highlight some of the recently proposed techniques to provide QoS continuity across the optical–wireless interface of FiWi access networks. For a more detailed and comprehensive description of QoS provisioning techniques for FiWi access networks the interested reader is referred to (Maier and Ghazisaidi [2010]).

17.4.1 QoS-aware scheduling

A centralized QoS-aware scheduling algorithm was proposed in (Luo *et al.* [2007]) to deliver services to both wireline and wireless users of an integrated EPON-WiMAX network. For QoS provisioning, the proposed scheduling algorithm utilizes the integrated ONU/WiMAX outdoor unit (ODU) to collect the resource information and service requests from all wireless subscribers. Whenever the wireless link conditions change, a WiMAX subscriber station sends a bandwidth request message to its associated ODU to update its service requirements. The collocated ONU aggregates the received bandwidth requests from multiple subscribers into a small number of multipoint control protocol (MPCP) REPORT messages and sends them upstream to the hybrid OLT/WiMAX base station, which acts as the central controller of the integrated EPON/WiMAX network. The aggregation of multiple bandwidth request messages into a small number of REPORT messages helps reduce the signaling overhead and achieve an improved throughput-delay performance. For each polling cycle, the central controller schedules the received bandwidth requests based on their traffic class and informs the WiMAX subscriber stations about the resultant upstream transmission schedule.

The presented simulation results show that the centralized QoS-aware scheduling algorithm is able to improve both network delay and throughput by roughly 10% compared with a distributed multihop scheduling algorithm, where each hop makes its own local transmission decision without any global knowledge of the status of all wireless links.

17.4.2 Enhanced MPCP and admission control

A resource management framework with the objective of maximizing the resource utilization and optimizing the QoS performance in both the optical and wireless segments of an integrated EPON/WiMAX network was investigated in (Yan *et al.* [2009]). The framework comprises two parts. The first part is the so-called *enhanced MPCP (E-MPCP)*, which aims at improving the signaling between EPON and WiMAX. The second part is called the *integrated optical wireless admission control (IOW-AC)* scheme, which is used to provide integrated bandwidth allocation in an EPON/WiMAX network.

In this approach, the collocated EPON ONU and WiMAX base station are integrated into a single device referred to as an access gateway. In E-MPCP, the GATE message is modified such that it contains an additional 3-byte field called the next cycle time, which denotes the time interval between two consecutive polling operations of a given access gateway. In doing so, the access gateway knows the estimated waiting time for its next poll and can thus calculate the transmission delay in the optical domain more accurately. The IOW-AC works as follows. Upon reception of a new service request, the access gateway calculates the transmission and propagation delays by taking the wireless network conditions and traffic profile into account. Subsequently, the access gateway calculates the queuing delay based on the current buffer occupancy. Finally, the access gateway also takes the aforementioned transmission delay in the optical segment into account and determines the estimated overall delay across both WiMAX and EPON networks. If the overall delay is smaller than a pre-specified delay bound the service request is accepted. Otherwise, the service request is rejected.

Simulation results were presented to compare the proposed integrated resource management framework with a conventional resource management scheme that controls the EPON and WiMAX networks separately from each other. The reported results show that E-MPCP and IOW-AC are superior to the separated resource management scheme in terms of channel utilization and dropping probability for both best-effort and real-time traffic. However, it is important to note that the performance of the integrated resource management framework largely depends on the network size and traffic profile.

17.4.3 QoS mapping

One important aspect of QoS continuity across the optical–wireless interface of FiWi access networks is the appropriate mapping of traffic classes of different optical and wireless broadband access technologies. In general, optical and wireless broadband technologies differ in the number and/or type of traffic classes. For instance, EPON deploys the eight priority traffic classes defined in IEEE standard 802.1D. On the other hand, we saw in Chapter 7 that fixed and mobile WiMAX defines four and five scheduling services respectively, to support different traffic classes, while in IEEE 802.11e QoS-enabled WLANs four different access categories are used, each associated with a different wireless channel access priority (see Chapter 6). These different priority queues, scheduling services, and access categories are used to meet the specific QoS requirements of different services and applications such as voice, video, email, web browsing, file transfer, etc. To provide end-to-end QoS support in FiWi access networks consisting of cascaded EPON, WiMAX, and/or WLAN networks for a variety of services and applications with different QoS requirements, the priority queues, scheduling services, and access categories must be mapped properly, which can be easily done in a straightforward manner (Lee *et al.* [2009]).

17.4.4 Optical burst wireless mesh architecture

A metropolitan area FiWi network architecture, called *optical burst wireless mesh architecture (OBWMA)*, which interconnects multiple wireless mesh networks (WMNs) at the user access side through an optical burst switching (OBS) core network, was proposed in (Belbekkouche *et al.* [2010, 2011], Rezgui *et al.* [2011]). One or more gateways of each WMN are connected to an OBS edge node. To provide end-to-end guaranteed QoS, OBWMA accepts a new flow request only when the flow end-to-end delay and bandwidth requirements can be met in the WMN. In the WMN, each mesh node has two radios with several orthogonal channels per radio. The wireless channels are statically assigned to each radio such that interference is minimized. The OBS core network is a wavelength division multiplexing (WDM) ring network consisting of several OBS edge nodes. Each OBS edge node aggregates received data packets into bursts based on the packets' destinations.

OBWMA supports two traffic classes. The high-priority traffic class has QoS requirements with respect to delay and bandwidth, while the low-priority traffic class doesn't have any QoS requirements (i.e., best-effort traffic). In the WMN, a certain percentage of the link capacity is allocated to each traffic class at every mesh node and either traffic class is allowed to exceed its allocated bandwidth only if the other traffic class has not reached its maximum bandwidth. Otherwise, excess traffic needs to be dropped. To enable loss-free transport of high-priority traffic across the OBS core network, each OBS network node is assigned a number of local wavelength channels that are dedicated to the transmission of high-priority traffic. Low-priority traffic is sent on any of the remaining contention-based wavelength channels or on the aforementioned local wavelength channels, provided that there is no high-priority traffic. To ensure end-to-end guaranteed QoS across the WMN and OBS networks, a control bridge is deployed at the OBS edge node, which has global knowledge about the OBWMA network resources. The control bridge performs QoS mapping between the WMN and OBS networks of OBWMA.

The performance of the OBWMA network was investigated in terms of throughput, delay, and packet loss by means of simulation. The obtained results show that at low traffic loads the end-to-end delay of the OBWMA network is dominated by the OBS core network. This is due to the fact that the burst assembly takes more time at low traffic loads. Conversely, at high traffic loads the WMN gets increasingly congested, resulting in an increased end-to-end delay. Furthermore, it was shown that the above described bandwidth allocation schemes help decrease the packet loss and increase the per-flow throughput of the network.

17.5 Summary

A number of routing algorithms covering only the wireless front-end of FiWi access networks have been studied with the goal of finding the optimal path by meeting one or more objectives. Pro-active link state delay-aware routing algorithms were proposed to minimize the average packet delay between any wireless mesh router and mesh gateway, improve load balancing, and provide QoS differentiation between delay-sensitive and delay-insensitive traffic. These new routing algorithms for the wireless mesh front-end of FiWi access networks were shown to outperform conventional routing algorithms such as minimum hop routing, shortest path routing, and predictive throughput routing in terms of delay, link congestion, and hop count. The delay-aware routing algorithms can be extended to distribute the radio capacity of a single-radio wireless mesh network node optimally among its outgoing links such that the average delay is minimized. To increase the capacity of the wireless front-end and better match it to the optical backhaul, a second radio might be placed at selected wireless routers and gateways. In multi-radio wireless mesh networks, channels must be carefully assigned such that traffic loads are balanced among wireless channels in order to minimize contention.

Risk-aware wireless routing algorithms can be developed to provide protection against multiple wireless front-end and/or optical back-end network failures, resulting in a significantly reduced packet loss in FiWi access networks.

A few integrated routing algorithms exist that combine some of the proposed wireless routing algorithms with the dynamic bandwidth allocation algorithms of a backhaul EPON with the objective of minimizing the end-to-end delay in FiWi access networks. In the design of integrated routing algorithms, it is key to take into account the fact that optical wired links are more reliable than their wireless counterparts. Toward this end, availability-aware routing algorithms were studied, which are able to achieve a higher average path availability than shortest path routing, at the expense of an increased mean number of hops. Another important criterion of integrated routing algorithms is their ability to put low-load ONUs to sleep in order to save energy. This can be achieved by designing energy-aware routing algorithms, which try to use already existent paths while putting other FiWi network segments to sleep. It must be noted, however, that energy-aware routing algorithms suffer from a fundamental trade-off between energy saving and QoS support. While several techniques exist to provide end-to-end guaranteed QoS in FiWi access networks, it remains an open challenge to yield both energy saving and QoS support in future green carrier-class FiWi broadband access networks.

18 Smart grid communications

The world has become heavily dependent on oil through the widespread use of combustion engines in gasoline cars, resulting in climate change, massive transfers of wealth to oil-producing countries, and heightened geopolitical tensions. The advent of commercially available *electric vehicles (EVs)* by the end of 2010 is expected to be a game changer that will shake things up in a fundamental manner. Not only can electricity be produced in a number of environmentally friendly ways, e.g., hydroelectric generators, wind farms, or solar arrays, but also the electric engine is significantly more efficient than the combustion engine of traditional gasoline cars or today's gasoline–electric hybrids (Davis [2010]). A promising example of using EVs for a sustainable electric mobility (e-mobility) in urban areas is the "e-mobility Berlin" project, which deploys only green electricity from renewable sources to realize a user-friendly public charging infrastructure for plug-in EVs (PEVs). The emission-free PEVs may be shared following DAIMLER's "car-to-go" idea and allow environmental zones to be set up in cities, from which environmentally friendly PEVs with no emissions are exempt (DAIMLER [2009]). Replacing gasoline vehicles with PEVs could reduce the importation of oil by up to 52% in the United States. Despite their huge potential to create new markets, revenues, and jobs, PEVs pose severe challenges to electric utility companies. A PEV being charged at home in the evening may more than double the average household electricity load and thereby dramatically exacerbates the load profile imbalance of power grids between off- and on-peak hours (Ipakchi and Albuyeh [2009]).

Today's power networks have to increase their utilization and become more efficient without depleting our ever declining natural resources to meet the increasing electric energy demand of a rapidly growing global population from 6.1 billion in 2000 to 7.5 billion by 2020, leading to a staggering 75% increase in power consumption by 2020 (Garrity [2008]). Toward this end, the current power grid has to be transformed into the so-called *smart grid* by incorporating sophisticated sensing, monitoring, information, and communications technologies to provide better grid performance and support a wide range of additional services to consumers. Massively deployed sensors will continuously collect data about end-user energy consumption, equipment condition, and operational status. The collected information will be disseminated through a highly available and flexible two-way communication infrastructure, also known as an advanced metering infrastructure (AMI), in order to predict the performance at any point of the power grid and develop optimal control strategies (Santacana *et al.* [2010]).

In this chapter, we first briefly review the vision of a smart grid and elaborate on its anticipated benefits, paying particular attention to the key role of homes. We then discuss short-term vs. long-term solutions to realize smart grid communications infrastructures and discuss the technological choices made to build a state-of-the-art smart grid communications infrastructure with currently available technologies, including lessons learned by pilot projects. We then identify a number of important requirements that future smart grid communications infrastructures have to meet and explain why fiber-wireless (FiWi) broadband access networks are more suitable to meet the aforementioned requirements of future smart grid communications solutions than alternative broadband access networks. Finally, we introduce a promising FiWi smart grid communications infrastructure, called the Über-FiWi network, and describe its architecture, implementation, and operation, and give a couple of highlighting results of its performance.

18.1 Smart grid vision

According to (Davies [2010]), no consensus has been reached on what the term smart grid exactly means. At present, there are too many technologies for the first/last mile access and home area and it is unclear which of them will finally prevail as the smart grid technologies of choice. The vision of the future smart grid is all about equipping intelligent devices with communication interfaces and engaging customers to play an interactive role in order to reduce power consumption and shift demand to minimize peak loads, which is widely referred to as *demand response*. The smart grid is sometimes called the *Energy Internet*, since its impact on our lifestyle and way of consuming and generating energy will be similar to how the Internet changed our way of communicating and sharing information (Davies [2010]).

The smart grid will be able to deliver not only energy but also information about every aspect of electricity generation and consumption bidirectionally, i.e., in both directions from and to each customer. The smart grid will be instrumental in addressing growing concerns about climate change and carbon gas emissions by exploiting renewable energy sources and controlling power consumption more efficiently, whereby homes will play a key role for the following reasons (Heile [2010]):

- Over 50% of the generated electricity is currently consumed in homes.
- PEVs will be charged mostly at homes after drivers return from work.
- Homes will be the location of the largest number of distributed power generation devices, e.g., wind turbines and solar panels, also known as *distributed energy resources (DERs)*.

Given the huge number of homes, smart grid solutions must be scalable and use access technologies that meet a number of specific requirements, as discussed in greater detail in the next section.

18.2 Smart grid communications infrastructure

In this section, we first briefly review Xcel Energy's approach to realizing its state-of-the-art *SmartGridCity* project in Boulder, Colorado, and elaborate on the rationale behind its technological choices. After discussing Xcel Energy's lessons learned from implementing a smart grid communications infrastructure with currently available technologies, we outline the requirements of emerging smart grid communications infrastructures and discuss a long-term solution that might evolve over the next couple of decades.

18.2.1 SmartGridCity Project

SmartGridCity[1] is one of the world's most advanced pilot projects that allows Xcel Energy to explore smart grid technologies and tools in a real-world setting. It went live in summer 2009 and connects nearly 47 000 premises throughout the city of Boulder.

Upon inquiry, Xcel Energy informed us that it utilizes broadband over power line (BPL) as the access technology of choice to connect to the individual premises and deploys more than 200 miles of fiber optic cable to backhaul the collected data. During the initial design of the system, Xcel Energy studied a number of alternatives to fiber, including satellite, cable, digital subscriber line (DSL), and wireless access technologies. The most promising alternative seemed to be utilizing a portion of the fiber backbone being installed by Qwest in Boulder as part of its ongoing DSL upgrade. In a thorough analysis, Xcel Energy reviewed the topology, type of fiber, availability of dark fiber, and location of fiber termination points of Qwest's build-out plan against the needs of the SmartGridCity project plan. The outcome of the analysis showed that the fiber from Qwest would be suitable for the project. However, the fiber termination points were not in close enough proximity to the project's injection points and the timing of the Qwest build-out was not sufficiently aligned to allow the SmartGridCity and Qwest systems to leverage on the same fiber backhaul. The other solutions considered in Xcel Energy's analysis were found to be unable to deal with underground or transformers and suffer from a number of different shortcomings. Specifically, satellite communications systems inherently exhibit high latency, wireless solutions have potential security and coverage issues and suffer from a lack of cards for existing meter types, while DSL falls short of providing complete coverage and ruggedized service offerings.

The SmartGridCity project provides invaluable insights into the opportunities and challenges of early smart grid deployments. Some of the lessons learned are of practical importance not only to Xcel Energy but also other electricity utilities intending to embark on smart grids. It is important to note, however, that in other parts of the United States and other regions of the world utilities might encounter different circumstances that shape their decision making process on which broadband access technologies to use

[1] Visit http://smartgridcity.xcelenergy.com for additional information about Xcel Energy's SmartGridCity project.

apart from BPL with its notorious difficulties (to be discussed in more detail shortly). Unlike Xcel Energy, other power companies might have a more flexible implementation schedule and prefer an evolutionary migration path for transforming their legacy power grid into the smart grid gradually over the next years or even decades by seizing new and future-proof technological opportunities and leveraging on a shared broadband access and smart grid communications infrastructure.

18.2.2 Requirements

Any viable smart grid communications solution must meet a number of important requirements. Beside scalability and coverage, possible smart grid communications infrastructures must provide security and easy maintenance given the fact that many of the existent utilities favor running their own network in order to achieve these two design objectives for backhaul communications (Heile [2010]). High-voltage substations in today's power grid measure large amounts of data, whereby the most fine-grained current and voltage measurements are sampled over 100 times a second (Bose [2010]). High-speed fiber links are well suited to carrying the voluminous amount of data gathered at a substation and transmit it to the smart grid control center across urban and regional areas. In addition, the supporting communications infrastructure must provide exceptionally tight latency characteristics and allow for real-time measurements with a maximum latency of 100 ms. While optical fiber is the preferred medium of choice in urban power distribution systems, WiMAX with its low latency performance may be used in less populated rural areas and locations, where fiber is not available or is available to some but not all points in the power distribution system. WiMAX supports quality-of-service (QoS) differentiation, which can be used to give control and protection traffic priority over other smart grid traffic types in order to render the power grid more survivable against failures that might lead to degraded service (brownout) or disrupted service (blackout) (Sood *et al.* [2009]). Despite the great debate whether WiMAX would become the dominant wireless broadband access technology and the fact that carriers are now going to bet on long term evolution (LTE) instead, WiMAX is expected to play a major role in new smart grid applications (Anderson [2010]). It is important to note, however, that the communications equipment must be much more robust in a smart grid than in carrier networks due to the fact that electricity substations and distribution grids are a much harsher environment than the telecom world (Davies [2010]). To cope with a variety of electromagnetic phenomena, WiMAX may be replaced with optical fiber networks, whose immunity against electromagnetic interference (EMI) makes them a superior choice. Similarly, wireless sensor networks (WSNs) have been widely recognized as a vital component of the smart grid to enable low-cost monitoring and control in order to maintain the safety, reliability, efficiency, and uptime of the smart grid. However, the harsh and complex electric power system environment poses great challenges to the reliability of WSN communications in smart grid applications (Gungor *et al.* [2010]). Furthermore, long-term smart grid communications solutions have to offer low capital and operational expenditures (CAPEX/OPEX), simple operation, administration, maintenance, and provisioning (OAM&P), longevity, future-proofness, as well as sustainability by means of a reduced CO_2 footprint and power consumption related to operation and heat dissipation.

FiWi networks hold great promise to help replace commuting with teleworking, which, taking the United States as an example, could lead to dramatic savings of 136 billion vehicle travel miles annually by 2020 and 171 billion miles by 2030 (Ruth [2009]). Previous work on FiWi networks focused only on the reduction of their own energy consumption and greenhouse gas emissions. It is only very recently that research has begun to study the role of access networks and adopt them also in other relevant sectors to enhance the efficiency of energy use, resulting in a dramatically improved overall CO_2 reduction across multiple sectors such as communications, energy, and transport (Somemura [2010]). Toward this end, in the next section we explore the opportunities of enhancing our SuperMAN network, described in Section 11.2.2, with fiber optical and wireless sensors and exploiting its unique characteristics to convert the traditional electric power grid, the largest man-created CO_2 emission source, into the future smart grid and to coordinate the interplay between PEVs and DERs at homes.

18.3 Über-FiWi network

In this section, we describe the architecture of the Über-FiWi network and discuss different implementation options that leverage on the benefits arising from the utilities' right-of-way and possible infrastructure sharing. We then explain the operation of the Über-FiWi network and evaluate its performance by means of simulation.

18.3.1 Architecture

Figure 18.1 depicts the architecture of the Über-FiWi network. A bidirectional fiber ring network is used to interconnect the distribution management system (DMS) with one or more WiMAX base stations (BSs) and optical line terminals (OLTs), whereby each OLT acts as the control unit of a separate attached Ethernet passive optical network (EPON). For survivability, the DMS, BSs, and OLTs might be interconnected via an additional fiber star subnetwork using dark fibers, which are readily available in metropolitan areas. The optical star subnetwork may deploy wavelength division multiplexing (WDM) to increase its capacity. The hub of the star subnetwork consists of a wavelength-broadcasting passive star coupler (PSC) and/or a wavelength-routing athermal arrayed-waveguide grating (AWG), depending on given capacity requirements and cost constraints. Both PSC and AWG are completely passive (i.e., unpowered) components, thus adding to the reliability, maintenance, and cost benefits of the proposed network architecture. The links of the star subnetwork can be used as back-up links in the event of link and/or node failures on the peripheral fiber ring. Alternatively, it can be used under normal (i.e., failure-free) operation to increase the capacity of the network and, more importantly, to provide short-cut links and thus help decrease the latency between subscribers and DMS. All links of

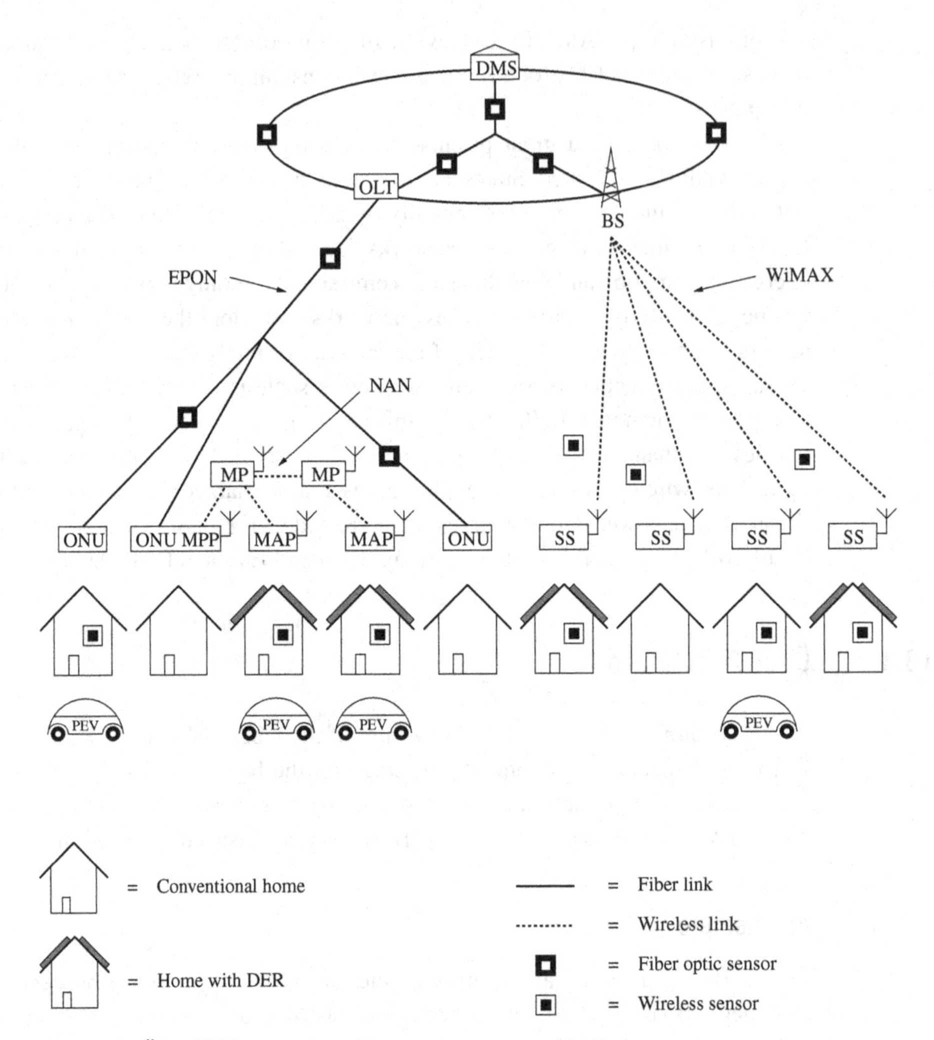

Figure 18.1 Über-FiWi network architecture.

the optical ring-star network architecture deploy simple and low-cost 1 Gb/s Ethernet technology.

Subscribers connect to the ring-star network either via EPON or WiMAX. The choice between these two broadband access technologies depends on the availability of fiber and a number of techno-economic factors. Recall from Chapter 13 that in terms of power consumption EPON is superior to WiMAX for different numbers of subscribers, terrain types, and mean access data rates. In terms of operational expenditures (OPEX), EPON is the preferred solution for deployment scenarios, where network failures are less likely. Taking both OPEX and capital expenditures (CAPEX) into account, WiMAX is more cost-efficient than EPON, especially in suburban and rural areas with a small population density. For densely populated settings, EPON represents a viable alternative solution that is able to be upgraded to higher data rates or multi-channel operation (WDM) without requiring any modifications of the installed fiber infrastructure.

As shown in Fig. 18.1, a subset of subscribers connect to the WiMAX BS via subscriber stations (SSs) wherever fiber is not yet available or economically prohibitive, e.g., rural areas. On the other hand, in densely populated or fiber-rich settings, subscribers connect to the EPON OLT via optical network units (ONUs). It is important to note that there is no need to connect each subscriber with a dedicated ONU. Instead, an ONU installed at a given subscriber's premises might be shared by other nearby subscribers through a WLAN-based mesh neighborhood area network (NAN), thereby significantly reducing the amount of required fiber infrastructure. The wireless mesh NAN consists of the following three types of network node: (*i*) mesh portal point (MPP), which is co-located at the ONU and acts as the interface to the NAN; (*ii*) mesh access points (MAPs), which provide subscribers with access to the NAN; and (*iii*) mesh points (MPs), which are the intermediate NAN nodes that relay packets between the MPP and MAPs. The NAN is based on high-throughput IEEE 802.11n WLAN technology with mesh support, as specified in IEEE standard 802.11s. Note that the NAN can also be used as a back-up network to re-route traffic between ONUs in case one ONU is disconnected from the OLT due to a fiber cut, leading to an improved EPON survivability.

To monitor the status of power distribution network components such as breakers, switches, voltage regulators, and transformers, we place fiber optic sensors, e.g., fiber Bragg grating (FBG) based temperature and voltage sensors, at the corresponding locations throughout the optical fiber infrastructure and wireless sensors, e.g., electric current and power sensors, in the WiMAX and NAN networks, as depicted in Fig. 18.1. The fiber optic sensors remotely communicate with interrogators located at the DMS using dedicated wavelength channels, while the electric sensors deploy the communications protocol defined in IEEE 802.15.4 ZigBee. In doing so, we transform our FiWi access network into a hybrid FiWi access-sensor network. Inside homes, we may deploy wireless sensors for enabling smart power metering and various home automation applications.

18.3.2 Implementation

Different approaches can be applied to implement the above described Über-FiWi network, depending on the preferred business model. A conservative business model might follow the traditional concept of *vertical integration*, whereby the network owner, network operator, and service provider are the same entity. Vertical integration has been the model applied by the vast majority of incumbents for decades. By applying this model, a utility could exploit its right-of-way and would leverage on the security and maintenance benefits of a private network, but would face significant investments to provide full coverage in its service area. According to (Breuer *et al.* [2011]), the future access network infrastructure business will be more divided in terms of ownership and operation, capitalizing on trust relationships between new alternative operators and incumbents. A promising approach to provide full coverage at reasonable costs is to jointly implement and share a common network infrastructure for both smart grid communications and broadband access, as discussed above in Section 18.2.1 for the case of Xcel Energy and Qwest. In such a business model, utilities, new (or incumbent) network operators, and service providers would team up and mutually benefit from each other by sharing costs related to the implementation and operation, administration, maintenance, and provisioning (OAM&P) of the Über-FiWi network. A business model that promotes the highest degree of competition and building an infrastructure for the society at large would be the so-called *open access network (OAN)* business model, where the roles of network owners, network operators, and service providers are separated (Forzati *et al.* [2010]). An OAN may be open at different levels, ranging from the infrastructure level to the service level. For instance, the infrastructure of the Über-FiWi network may be installed and owned by one or more power utilities,[2] run by a new or incumbent network operator, and may support a variety of service providers from different sectors, e.g., e-health, whereby the service providers are responsible for interacting with customers and addressing issues related to billing, authentication, authorization, privacy, and security (Battiti *et al.* [2005]).

18.3.3 Operation

The optical and wireless technologies in the Über-FiWi network offer some inherent mechanisms for QoS support and service differentiation. Specifically, IEEE 802.16 specifies five scheduling services for WiMAX, whereby the so-called unsolicited grant service (UGS) provides absolute QoS for periodic high-priority traffic. UGS is also well suited to carrying periodic measurements of wireless sensors across WiMAX networks. In the wireless mesh NAN, enhanced distributed channel access (EDCA), specified in IEEE 802.11e, and mesh deterministic access (MDA), specified in IEEE 802.11s, can be applied to provide relative and absolute QoS, respectively. Among others, MDA may be used to transfer periodic traffic originating from wireless sensors across the NAN.

The above described Über-FiWi network and access control protocols are used not only to provide end-users with triple-play services but also to enable subscribers residing in different homes to exchange information about their power usage. In doing so, power consumption across neighboring houses can be coordinated to reduce the peak demand by means of *inter-home scheduling* algorithms. It was recently shown in (Kishore and Snyder [2010]) that if each individual home optimizes its power consumption by taking advantage of off-peak prices independently from other homes (i.e., intra-home scheduling), the problem of demand peaks is not alleviated, but rather exaggerated, leading to a new "rebound" peak right after the peak hours. This "rebound" peak may be larger than the original peak, thereby negating some of the benefits of off-peak pricing models. The authors proposed a distributed inter-home scheduling algorithm among neighboring homes, which controls the usage of appliances across

[2] For instance, *openaxs* (http://www.openaxs.ch) is a consortium of Swiss power utilities that develops fiber-to-the-home (FTTH)-based OANs.

multiple homes and helps utilities effectively reduce the peakiness of their demand load. However, no specific inter-home communications infrastructure or access control protocol were presented in (Kishore and Snyder [2010]). Furthermore, the Über-FiWi network enables subscribers to exchange information about the presence and status of DERs and PEVs at selected homes within the same neighborhood, as shown in Fig. 18.1. DERs may be connected to the power grid according to the requirements specified in IEEE 1547. The information exchanged among subscribers through the Über-FiWi network also takes into account the existence of additional electric power lines between neighboring homes, if any, which were proposed in (Yoneda [2010]) to realize low-carbon power infrastructures. These power lines would allow power to be directly transferred between DERs and PEVs located at different homes, whereby PEVs might be also used as distributed energy storages. More interestingly, these power lines would also allow the complementary nature of intermittent renewable energy sources to be capitalized on, e.g., wind turbines and solar panels, and delay-tolerant PEVs, whose battery charging can be flexibly scheduled between evening and morning, as investigated in more detail in the following.

18.3.4 Performance

The province of Ontario, Canada, has been actively pursuing the deployment of smart meters and electricity time-of-use pricing with different off-peak, mid-peak, and on-peak prices for weekdays for summer and winter, as shown in Fig. 18.2.

In the following, we investigate the total power consumption and cost for 50 homes, each running a dishwasher, clothes dryer, and water heater assuming the appliance parameters of Table I in (Kishore and Snyder [2010]). In our simulations, 32 homes deploy their own ONU to connect to the OLT while the remaining 18 homes uniformly access the ONUs via the wireless mesh NAN. Figure 18.3 depicts the total power consumption for a summer weekday with and without intra-home scheduling as well as with inter-home scheduling among the 50 homes. The total power consumption varies with the time of day without any major demand peaks following the random appliance requests. In contrast, with intra-home scheduling, appliances in each separate

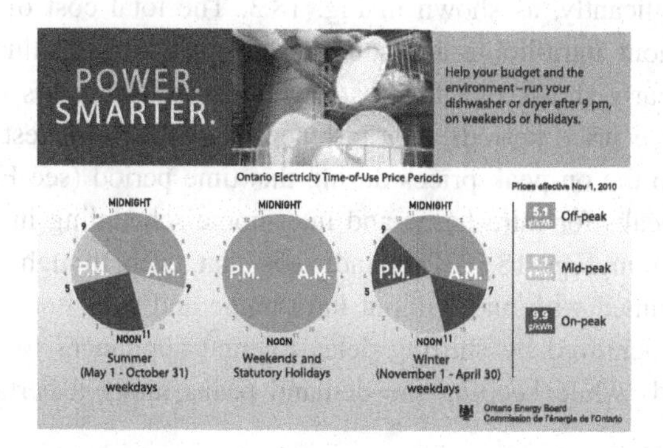

Figure 18.2 Ontario electricity time-of-use price periods. ©2010 Ontario Energy Board.

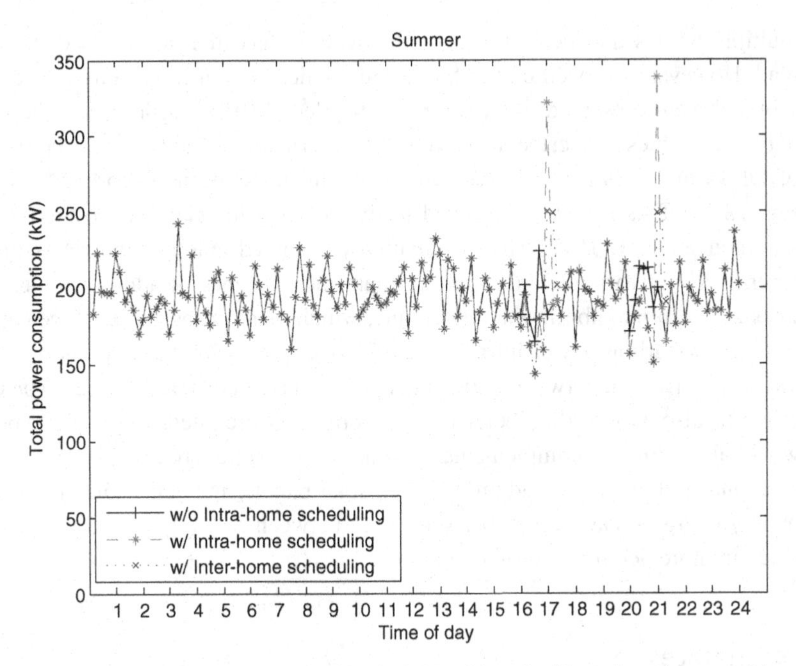

Figure 18.3 Total power consumption (in kW) with and without intra-home and inter-home scheduling.

home might be delayed by up to 60 minutes in order to benefit from mid-peak or off-peak prices. We observe from Fig. 18.3 that such a cost-saving behavior leads to severe power demand peaks at 5 p.m. (transition from on-peak to mid-peak price) and 9 p.m. (transition from mid-peak to off-peak price). To alleviate this problem, we deploy a centralized inter-home scheduling algorithm at the OLT that limits the maximum total power consumption while meeting the delay tolerance of appliances. Specifically, the OLT collects the appliance requests from all homes via the Über-FiWi network and then schedules the requests in a first-come-first-served manner such that no appliance is delayed by more than 60 minutes. In doing so, the power demand peaks can be easily upper limited according to given utility preferences without increasing the delay significantly, as shown in Fig. 18.3. The total cost of consuming power with and without intra-home and inter-home scheduling is illustrated in Fig. 18.4. The figures clearly shows that the cost increases and decreases according to the different time-of-use price periods. For instance, the cost is highest between 11 a.m. and 5 p.m. due to the on-peak prices during this time period (see Fig. 18.2). Note, however, that the peaks for intra-home and inter-home scheduling in Fig. 18.4 are less pronounced than in Fig. 18.3. This indicates that even though the same amount of power is consumed with and without intra-home and inter-home scheduling, cost savings can be achieved by shifting delay-tolerant appliances to mid-peak and off-peak price periods while keeping the demand peaks under a certain upper limit.

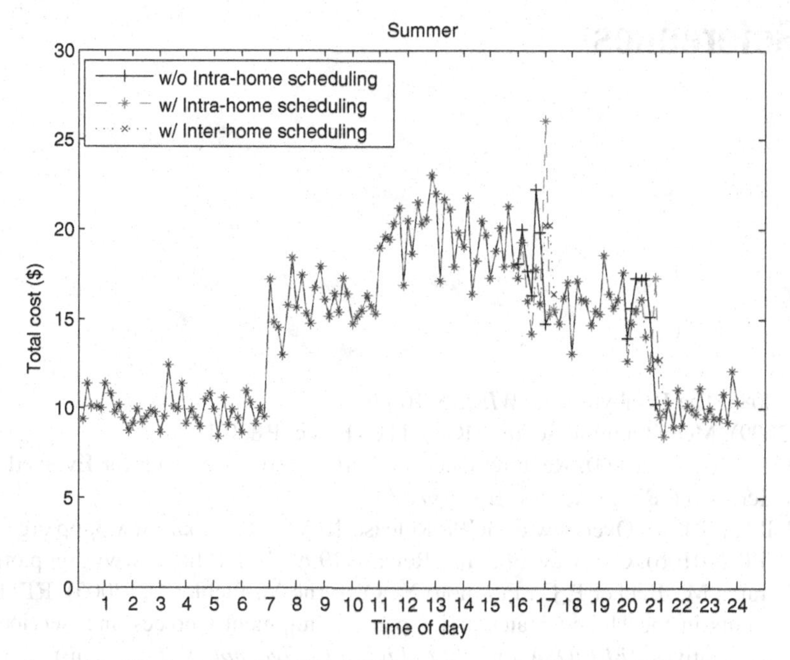

Figure 18.4 Total cost (in CAN$) with and without intra-home and inter-home scheduling.

18.4 Summary

Apart from reducing their own energy consumption and greenhouse gas emissions, FiWi access networks may also be adopted in economic sectors other than communications such as energy and transportation in order to enhance the efficiency of energy use and achieve an increased overall CO_2 reduction across different sectors. In this chapter, we explored the opportunities of deploying FiWi broadband access networks enhanced with fiber optic and wireless sensors to realize a smart grid communications infrastructure that effectively helps reduce the demand peakiness of today's power grid by means of centralized inter-home scheduling for electricity time-of-use pricing, which is an important ingredient of the future smart grid.

References

(2008). The Petabyte Age. *WIRED*, 106 ff.

(2009). Mean Opinion Score (MOS). ITU-T Rec. P.800.1.

36.913 G. T. (2009). Requirements for Further Advancements for Evolved-Universal Terrestrial Radio Access (E-UTRA).

3GPP. (2010a). Overview of 3GPP Release 10 V0.0.8. http://www.3gpp.org.

3GPP. (2010b). Overview of 3GPP Release 1999 V0.1.1. http://www.3gpp.org.

Abrams M., Becker P. C., Fujimoto Y., O'Byrne V., Piehler D. (2005). FTTP Deployments in the United States and Japan – Equipment Choices and Service Provider Imperatives. *IEEE/OSA Journal of Lightwave Technology 23*, 1 (Jan.), 236–246.

Abu Ali N. A., Taha A.-E. M., Hassanein H. S., Mouftah H. T. (2008). IEEE 802.16 Mesh Schedulers: Issues and Design Challenges. *IEEE Network 22*, 1 (Jan./Feb.), 58–65.

Ahmadi S. (2009). An Overview of Next-Generation Mobile WiMAX Technology. *IEEE Communications Magazine 47*, 6 (June), 84–98.

Aissa S., Maier M. (2007). Towards Seamless Fiber-Wireless (FiWi) Access Networks: Convergence and Challenges (Invited Paper). In *Proc., International Conference on Transparent Optical Networks (ICTON)–'Mediterranean Winter' 2007*. 1–6.

Akyildiz I., Wang X., Wang W. (2005). Wireless Mesh Networks: A Survey. *Computer Networks 47*, 4 (Mar.), 445–487.

Ali M. A., Ellinas G., Erkan H., Hadjiantonis A., Dorsinville R. (2010). On the Vision of Complete Fixed-Mobile Convergence. *IEEE/OSA Journal of Lightwave Technology 28*, 16 (Aug.), 2343–2357.

Anas M., Rosa C., Calabrese F., Pedersen K., Mogensen P. (2008). Combined Admission Control and Scheduling for QoS Differentiation in LTE Uplink. In *Proc., IEEE Vehicular Technology Conference (VTC)*. 1–5.

Anderson M. (2010). WiMAX for Smart Grids. *IEEE Spectrum 47*, 7 (July), 14.

Assi C. M., Ye Y., Dixit S., Ali M. A. (2003). Dynamic Bandwidth Allocation for Quality-of-Service over Ethernet PONs. *IEEE Journal on Selected Areas in Communications 21*, 9 (Nov.), 1467–1477.

Astely D., Dahlman E., Furuskar A., *et al.* (2009). LTE: The Evolution of Mobile Broadband. *IEEE Communications Magazine 47*, 4 (Apr.), 44–51.

Bahr M. (2006). Proposed Routing for IEEE 802.11s WLAN Mesh Networks. In *Proc., International Workshop on Wireless Internet (WICON)*. 1–10.

Bahr M. (2007). Update on the Hybrid Wireless Mesh Protocol of IEEE 802.11s. In *Proc., IEEE Internatonal Conference on Mobile Adhoc and Sensor Systems (MASS)*. 1–6.

Bahr M. (2009). IEEE P802.11 Wireless LANs: Resolution to CID 1736 – HWMP Overview, IEEE P802.11 - TASK GROUP S.

Baliga J., Ayre R., Sorin W. V., Hinton K., Tucker R. S. (2008). Energy Consumption in Access Networks. In *Proc., Optical Fiber Communication/National Fiber Optic Engineers Conference (OFC/NFOEC)*. 1–3.

Battiti R., Cigno R. L., Sabel M., Orava F., Pehrson B. (2005). Wireless LANs: From WarChalking to Open Access Networks. *Mobile Networks and Applications 10*, 3 (Aug.), 275–287.

Belbekkouche A., Rezgui J., Hafid A. (2010). QoS provisioning for Wireless Mesh and Optical Burst Switching Convergence. In *Proc., IEEE Wireless Communications and Networking Conference (WCNC)*. 1–6.

Belbekkouche A., Rezgui J., Hafid A. (2011). Wireless Mesh and Optical Burst Switching Convergence for a Novel Metropolitan Area Network Architecture. *Computer Networks 55*, 1 (Jan.), 159–172.

Belzner M., Haunstein H. (2009). Network Coding in Passive Optical Networks. In *Proc., European Conference on Optical Communication (ECOC)*. 1–2.

Bem D. J., Wieckowski T. W., Zielinski R. J. (2000). Broadband Satellite Systems. *IEEE Communications Surveys & Tutorials 3*, 1 (First Quarter), 2–15.

Bertsch L. A. (1990). Development Tools for Home Automation. *IEEE Transactions on Consumer Electronics 36*, 4 (Nov.), 854–858.

Bettstetter C., Vogel H.-J., Eberspacher J. (1999). GSM Phase 2+; General Packet Radio Service GPRS: Architecture, Protocols, and Air Interface. *IEEE Communications Surveys & Tutorials 2*, 3 (Third Quarter), 2–14.

Bhandari S., Park E. K. (2006). Hybrid Optical Wireless Networks. In *Proc., IEEE International Conference on Networking, International Conference on Systems and International Conference on Mobile Communications and Learning Technologies (ICN/ICONS/MCL)*. 1–5.

Biagini W., Yang N. (2000). Wireless Multimedia Communication through Multichannel Multipoint Distribution Service. In *Proc., International Conference on Information Technology: Coding and Computing (ITCC)*. 98–103.

Blake V. (2008). DOCSIS over PON. In *Proc., Optical Fiber Communication/National Fiber Optic Engineers Conference (OFC/NFOEC), paper OThT1*.

Bontu C., Illidge E. (2009). DRX Mechanism for Power Saving in LTE. *IEEE Communications Magazine 47*, 6 (June), 48–55.

Bose A. (2010). Smart Transmission Grid Applications and Their Supporting Infrastructure. *IEEE Transactions on Smart Grid 1*, 1 (June), 11–19.

Breuer D., Geilhardt F., Hulsermann F. *et al.* (2011). Opportunities for Next-Generation Optical Access. *IEEE Communications Magazine, Special Issue on Advances in Passive Optical Networks 49*, 2 (Feb.), S16–S24.

Byun H.-J., Nho J.-M., Lim J.-T. (2003). Dynamic Bandwidth Allocation Algorithm in Ethernet Passive Networks. *IEE Electronics Letters 39*, 13 (June), 1001–1002.

CableLabs. (2010). http://www.cablelabs.com.

Cai J., Goodman D. J. (1997). General Packet Radio Service in GSM. *IEEE Communications Magazine 35*, 10 (Oct.), 122–131.

Callendar M. H. (2010). International Mobile Telecommunications (IMT): Cellular and Broadband Access for the 21st Century. http://www.itu.int.

Camp J., Knightly E. (2008). The IEEE 802.11s Extended Service Set Mesh Networking Standard. *IEEE Communications Magazine 46*, 8 (Aug.), 120–126.

Campanella S. J., Harrington J. V. (1984). Satellite Communications Networks. *Proceedings of the IEEE 72*, 11 (Nov.), 1506–1519.

Carr N. (2008). *The Big Switch: Rewiring the World, from Edison to Google*. W. W. Norton & Company, Inc.

Chan V. W. S. (2003). Optical Satellite Networks. *IEEE/OSA Journal of Lightwave Technology 21*, 11 (Nov.), 2811–2827.

Chang C.-H., Kourtessis P., Senior J. M. (2006). GPON Service Level Agreement Based Dynamic Bandwidth Assignment Protocol. *IET Electronics Letters 42,* 20 (Sept.), 1173–1174.

Chang G.-K., Jian W., Jia Z., Chowdhury A. (2010). Broadband Access Technologies for Very High Throughput Wireless Sensor Communication Networks. In *Proc., IEEE Radio and Wireless Symposium (RWS)*. 496–499.

Chen J.-C., Chen W.-M. (2007). Design and Analysis of a Mobility Gateway for GPRS–WLAN Integration. *IEEE Transactions on Vehicular Technology 56*, 5 (Sept.), 2603–2616.

Chen S., Setta M., Chen X., Parini C. G. (2009). Ultra Wideband Powerline Communication (PLC) above 30 MHz. *IET Communications 3*, 10, 1587–1596.

Chen X., Reaz A., Shi L. *et al.* (2010). Delay-Differentiated Routing Algorithm to Enhance Delay Performance of WOBAN. In *Proc., International Conference on Optical Internet (COIN)*. 1–4.

Cheng H. T., Zhuang W. (2009). QoS-Driven MAC-Layer Resource Allocation for Wireless Mesh Networks with Non-Altruistic Node Cooperation and Service Differentiation. *IEEE Transactions on Wireless Communications 8*, 12 (Dec.), 6089–6103.

Cho K. Y., Takushima Y., Chung Y. C. (2008). 10-Gb/s Operation of RSOA for WDM PON. *IEEE Photonics Technology Letters 20,* 18 (Sept.), 1533–1535.

Choi S.-I., Huh J.-D. (2002). Dynamic Bandwidth Allocation Algorithm for Multimedia Services over Ethernet PONs. *ETRI Journal 24*, 6 (Dec.), 465–468.

Chowdhury A., Chien H.-C., Hsueh Y.-T., Chang G.-K. (2009a). Advanced System Technologies and Field Demonstration for In-Building Optical-Wireless Network with Integrated Broadband Services. *IEEE/OSA Journal of Lightwave Technology 27*, 12 (June), 1920–1927.

Chowdhury P., Mukherjee B., Sarkar S., Kramer G., Dixit S. (2009b). Hybrid Wireless-Optical Broadband Access Network (WOBAN): Prototype Development and Research Challenges. *IEEE Network 23*, 3 (May/June), 41–48.

Chowdhury P., Sarkar S., Reaz A. A. (2008). Comparative Cost Study of Broadband Access Technologies. In *Proc., International Symposium on Advanced Networks and Telecommunication Systems (ANTS)*. 1–3.

Chowdhury P., Tornatore M., Sarkar S., Mukherjee B. (2009c). Towards Green Broadband Access Networks. In *Proc., IEEE GLOBECOM*. 1–6.

Chowdhury P., Tornatore M., Sarkar S., Mukherjee B. (2010). Building a Green Wireless-Optical Broadband Access Network (WOBAN). *IEEE/OSA Journal of Lightwave Technology 28*, 16 (Aug.), 2219–2229.

Chu T.-S., Gans M. J. (1991). Fiber Optic Microcellular Radio. *IEEE Transactions on Vehicular Technology 40*, 3 (Aug.), 599–606.

Cicconetti C., Lenzini L., Mingozzi E., Eklund C. (2006). Quality of Service Support in IEEE 802.16 Networks. *IEEE Network 20*, 2, 50–55.

Ciciora W. S. (2001). The Cable Modem Traffic Jam. *IEEE Spectrum 38*, 6 (June), 48–53.

Cioffi J. M., Jagannathan S., Mohseni M., Ginis G. (2007). CuPON: The Copper Alternative to PON 100 Gb/s DSL Networks. *IEEE Communications Magazine 45*, 6 (June), 132–139.

Cioffi J. M., Oksman V., Werner J.-J. *et al.* (1999). Very-High-Speed Digital Subscriber Lines. *IEEE Communications Magazine 37*, 4 (Apr.), 72–79.

Clark D. (1998). Powerline Communications: Finally Ready for Prime Time? *IEEE Internet Computing 2*, 1 (Jan./Feb.), 10–11.

Correia N., Coimbra J., Schütz G. (2009). Fault-Tolerance Planning in Multiradio Hybrid Wireless-Optical Broadband Access Networks. *IEEE/OSA Journal of Optical Communications and Networking 1*, 7 (Dec.), 645–654.

Cvijetic N., Qian D., Hu J., Wang T. (2010). Orthogonal Frequency Division Multiple Access PON (OFDMA-PON) for Colorless Upstream Transmission Beyond 10 Gb/s. *IEEE Journal on Selected Areas in Communications 28*, 6 (Aug.), 781–790.

Czajkowski I. K. (1999). High-Speed Copper Access: A Tutorial Overview. *Electronics & Communication Engineering Journal 11*, 3 (June), 125–148.

DAIMLER. (2009). 'E-mobility Berlin' Delivers Smart Solutions for Electric Mobility. *White Paper*.

Dang B. L., Niemegeers I. (2005). Analysis of IEEE 802.11 in Radio over Fiber Home Networks. In *Proc., IEEE Conference on Local Computer Networks*. 744–747.

Das A., Nkansah A., Gomes N. J., *et al.* (2006). Design of Low-Cost Multimode Fiber-Fed Indoor Wireless Networks. *IEEE Transactions on Microwave Theory and Techniques 54*, 8 (Aug.), 3426–3432.

Davies S. (2010). Grid Looks to Smart Solutions. *IET Engineering & Technology 5*, 7 (May), 49–51.

Davik F., Yilmaz M., Gjessing S., Uzun N. (2004). IEEE 802.17 Resilient Packet Ring Tutorial. *IEEE Communications Magazine 42*, 3 (Mar.), 112–118.

Davis C. C., Smolyaninov I. I., Millner S. D. (2003). Flexible Optical Wireless Links and Networks. *IEEE Communications Magazine 41*, 3 (Mar.), 51–57.

Davis J. (2010). Supercharged. *WIRED*, 138–145,176–180.

Demir C., Comaniciu C. (2007). An Auction based AODV Protocol for Mobile Ad Hoc Networks with Selfish Nodes. In *Proc., IEEE International Conference on Communications (ICC)*. 3351–3356.

Diallo A., Luxey C., Le Thuc P., Staraj R., Kossiavas G. (2008). Enhanced Two-Antenna Structures for Universal Mobile Telecommunications System Diversity Terminals. *IET Microwaves Antennas Propagation 2*, 1 (Feb.), 93–101.

Donaldson G., Jones D. (2001). Cable Television Broadband Network Architectures. *IEEE Communications Magazine 39*, 6 (June), 122–126.

Draves R., Padhye J., Zill B. (2004). Routing in Multi-Radio, Multi-Hop Wireless Mesh Networks. In *Proc., ACM MobiCom*. 114–128.

Dulac S. P., Godwin J. P. (2006). Satellite Direct-to-Home. *Proceedings of the IEEE 94*, 1 (Jan.), 158–172.

Dutta-Roy A. (1999). Cable: It's Not Just for TV. *IEEE Spectrum 36*, 5 (May), 53–59.

Dutta-Roy A. (2001). An Overview of Cable Modem Technology and Market Perspectives. *IEEE Communications Magazine 39*, 6 (June), 81–88.

Effenberger F., Clearly D., Haran O. *et al.* (2007). An Introduction to PON Technologies. *IEEE Communications Magazine 45*, 3 (Mar.), S17–S25.

Effenberger F., Mukai H., Park S., Pfeiffer T. (2009a). Next-Generation PON – Part II: Candidate Systems for Next-Generation PON. *IEEE Communications Magazine 47*, 11 (Nov.), 50–57.

Effenberger F. J., Mukai H., Kani J.-I., Rasztovits-Wiech M. (2009b). Next-Generation PON – Part III: System Specifications for XG-PON. *IEEE Communications Magazine 47*, 11 (Nov.), 58–64.

Effros M., Ho T., Kim S. (2006). A Tiling Approach to Network Code Design for Wireless Networks. In *Proc., Information Theory Workshop*. 62–66.

Egeland G., Engelstad P. (2009). The Availability and Reliability of Wireless Multi-Hop Networks with Stochastic Link Failures. *IEEE Journal on Selected Areas in Communications 27*, 7 (Sept.), 1132–1146.

Eklund C., Marks R. B., Stanwood K. L., Wang S. (2002). IEEE Standard 802.16: A Technical Overview of the WirelessMAN Air Interface for Broadband Wireless Access. *IEEE Communications Magazine 40*, 6 (June), 98–107.

Eldin S. S., Nasr M., Khamees S., Sourour E., Elbanna M. (2009). Performance Enhancement of IEEE 802.11n Wireless LAN Using Irregular LDPCC. In *Proc., IEEE/IFIP International Conference on Wireless and Optical Communications Networks (WOCN)*. 1–5.

Epple B., Henniger H. (2007). Discussion on Design Aspects for Free-Space Optical Communication Terminals. *IEEE Communications Magazine 45*, 10 (Oct.), 62–69.

Ericsson. (2008). Deep Fiber Access. *White Paper*.

Eryilmaz A., Ozdaglar A., Médard M. (2006). On Delay Performance Gains from Network Coding. In *Proc., CISS*. 864–870.

Espes D., Mammeri Z. (2007). Improvement of AODV Routing in Dense Networks. In *Proc., IEEE International Symposium on a World of Wireless, Mobile and Multimedia Networks (WoWMoM)*. 1–4.

Etemad K. (2008). Overview of Mobile WiMAX Technology and Evolution. *IEEE Communications Magazine 46*, 10 (Oct.), 31–40.

Faccin S., Wijting C., Kenckt J., Damle A. (2006). Mesh WLAN Networks: Concept and System Design. *IEEE Wireless Communications 13*, 2 (Apr.), 10–17.

Fan K., Wei X., Long D. (2009). A Load-Balanced Route Selection for Network Coding in Wireless Mesh Networks. In *Proc., IEEE International Conference on Communications (ICC)*. 1–6.

Federal Communications Commission. (2010). Sixth Broadband Deployment Report. *FCC 10-129*.

Fellows D., Jones D. (2001). DOCSIS Cable Modem Technology. *IEEE Communications Magazine 39*, 3 (Mar.), 202–209.

Feng T., Ruan L. (2009). Design of Survivable Hybrid Wireless-Optical Broadband-Access Network. In *Proc., IEEE International Conference on Communications (ICC)*. 1–5.

Foh C. H., Andrew L., Wong E., Zukerman M. (2004). FULL-RCMA: A High Utilization EPON. *IEEE Journal on Selected Areas in Communications 22*, 8 (Oct.), 1514–1524.

Forzati M., Larsen C. P., Mattsson C. (2010). Open Access Networks, the Swedish Experience. In *Proc., International Conference on Transparent Optical Networks (ICTON)*. 1–4.

Fouli K., Berisa T., Maier M. (2009). Optical Coding for Enhanced Real-Time Dynamic Bandwidth Allocation in Passive Optical Networks. *IEEE/OSA Journal of Lightwave Technology 27*, 23 (Dec.), 5376–5384.

Fouli K., Maier M. (2007). OCDMA and Optical Coding: Principles, Applications, and Challenges. *IEEE Communications Magazine 45*, 8 (Aug.), 27–34.

Fouli K., Maier M., Médard M. (2011). Network Coding in Next-Generation Passive Optical Networks. *IEEE Communications Magazine 49*, 9 (Sep.), 38–46.

Fragouli C., Katabi D., Markopoulou A., Médard M., Rahul H. (2007). Wireless Network Coding: Opportunities & Challenges. In *Proc., IEEE MILCOM*. 1–8.

Fu I.-K., Chen Y.-S., Cheng P., *et al.* (2010). Multicarrier Technology for 4G WiMax System. *IEEE Communications Magazine 48*, 8 (Aug.), 50–58.

Furuskär A., Mazur S., Müller F., Olofsson H. (1999). EDGE: Enhanced Data Rates for GSM and TDMA/136 Evolution. *IEEE Personal Communications 6*, 3 (June), 56–66.

Garrity T. F. (2008). Getting Smart. *IEEE Power & Energy Magazine 6*, 2 (March/April), 38–45.

Garroppo R., Giordano S., Iacono D., Tavanti L. (2008). Notes on Implementing a IEEE 802.11s Mesh Point. In *Proc., EuroNGI Workshop*. 60–72.

Ge Y., Wen S., Ang Y.-H., Liang Y.-C. (2010). Optimal Relay Selection in IEEE 802.16j Multihop Relay Vehicular Networks. *IEEE Transactions on Vehicular Technology 59*, 5 (June), 2198–2206.

Gershon R., Propp D., Propp M. (1991). A Token Passing Network for Powerline Communications. *IEEE Transactions on Consumer Electronics 37*, 2 (May), 129–134.

Ghazisaidi N., Maier M. (2010). Techno-Economic Analysis of EPON and WiMAX for Future Fiber-Wireless (FiWi) Networks. *Computer Networks 54*, 15 (Oct.), 2640–2650.

Ghazisaidi N., Maier M. (2011). Hierarchical Frame Aggregation Techniques for Hybrid Fiber-Wireless Access Networks. *IEEE Communications Magazine 49*, 9 (Sep.), 64–73.

Ghazisaidi N., Paolucci F., Maier M. (2009). SuperMAN: Optical-Wireless Integration of RPR and WiMAX. *OSA Journal of Optical Networking 8*, 3 (Mar.), 249–271.

Ghazisaidi N., Scheutzow M., Maier M. (2010). Frame Aggregation in Fiber-Wireless (FiWi) Broadband Access Networks. *IET Electronics Letters 46*, 5 (Mar.), 377–379.

Ghazisaidi N., Scheutzow M., Maier M. (2011). Survivability Analysis of Next-Generation Passive Optical Networks and Fiber-Wireless Access Networks. *IEEE Transactions on Reliability 60*, 2 (June), 479–492.

Gladisch A., Lange C., Leppla R. (2008). Power Efficiency of Optical Versus Electronic Access Networks (Invited Paper). In *Proc., European Conference on Optical Communication (ECOC)*.

Gong M., Lin B., Ho P.-H. (2010). BU Association and Resource Allocation in Integrated PON-WiMAX under Inter-cell Cooperative Transmission. In *Proc., IEEE International Symposium on Wireless Pervasive Computing (ISWPC)*. 606–611.

Green P. E. (2006). *Fiber To The Home – The New Empowerment*. John Wiley & Sons, Inc.

Green R. J., Joshi H., Higgins M. D., Leeson M. S. (2008). Recent Developments in Indoor Optical Wireless Systems. *IET Communications 2*, 1 (Jan.), 3–10.

Grodzinsky M., de Vegt R. (2008). Introduction to VHT Usage Models.

Guezouri M., Ouamri A. (2007). Optimizing Routes Quality and Scattering in the AODV Routing Protocol. *Journal of Computer Science & Technology 7*, 3 (Oct.), 209–212.

Gungor V. C., Lu B., Hancke G. P. (2010). Opportunities and Challenges of Wireless Sensor Networks in Smart Grid – A Case Study of Link Quality Assessments in Power Distribution Systems. *IEEE Transactions on Industrial Electronics 57*, 10 (Oct.), 3557–3564.

Haas Z. J., Pearlman M. R., Samar P. (2002). The Zone Routing Protocol (ZRP) for Ad Hoc Networks. IETF Internet-Draft (July).

Hajduczenia M., da Silva H. J. A., Monteiro P. P. (2006). EPON versus APON and GPON: A Detailed Performance Evaluation. *OSA Journal of Optical Networking 5*, 4 (Apr.), 298–319.

Hajduczenia M., Silva H. J. A. D., Monteiro P. (2008). Development of 10 Gb/s EPON in IEEE 802.3av. *IEEE Communications Magazine 46*, 7 (July), 40–47.

Han J., Wu B. (2010). Handover in the 3GPP Long Term Evolution (LTE) Systems. In *Proc., Global Mobile Congress (GMC)*. 1–6.

Handler D. P., Grossman R. (2009). The Role of Income Distribution and Broadband Penetration in Developing Countries. *Cisco White Paper*.

Harrison F. G., Hearnden S. R. (1999). The Challenge to Realise Convergence of Fixed and Mobile Communications. *Electronics & Communication Engineering Journal 11*, 3 (June), 164–168.

Hazen M. E. (2008). The Technology Behind HomePlug AV Powerline Communications. *IEEE Computer 41*, 6 (June), 90–92.

He T., Chan S.-H., Wong C.-F. (2008). HomeMesh: A Low-Cost Indoor Wireless Mesh for Home Networking. *IEEE Communications Magazine 46*, 12 (Dec.), 79–85.

Heile B. (2010). Smart Grids for Green Communications. *IEEE Wireless Communications 17*, 3 (June), 4–6.

Henry P. S. (2007). Integrated Optical/Wireless Alternatives for the Metropolitan Environment. *IEEE Communications Society Webinar*.

Herzog M., Maier M. (2006). RINGOSTAR: An Evolutionary Performance-Enhancing WDM Upgrade of IEEE 802.17 Resilient Packet Ring. *IEEE Communications Magazine 44*, 2 (Feb.), S11–S17.

Hiertz G., Zang Y., Max S., *et al.* (2008). IEEE 802.11s: WLAN Mesh Standardization and High Performance Extensions. *IEEE Network 22*, 3 (May/June), 12–19.

Ho T., Lun D. (2008). *Network Coding: An Introduction*. Cambridge University Press.

Holma H., Toskala A., Ranta-aho K., Pirskanen J. (2007). High-Speed Packet Access Evolution in 3GPP Release 7. *IEEE Communications Magazine 45*, 12 (Dec.), 29–35.

HomePlug Powerline Alliance. (2010). http://www.homeplug.org.

Hong C.-Y., Pang A.-C. (2009). 3-Approximation Algorithm for Joint Routing and Link Scheduling in Wireless Relay Networks. *IEEE Transactions on Wireless Communications 8*, 2 (Feb.), 856–861.

Horrigan J. (2009). Home Broadband Adoption 2009. *Pew Internet & American Life Project*.

Hu Y., Li V. O. K. (2001). Satellite-Based Internet: A Tutorial. *IEEE Communications Magazine 39*, 3 (Mar.), 154–162.

Huang J., Subramanian V., Agrawal R., Berry R. (2009). Joint Scheduling and Resource Allocation in Uplink OFDM Systems for Broadband Wireless Access Networks. *IEEE Journal on Selected Areas in Communications 27*, 2 (Feb.), 226–234.

Humphrey M., Freeman J. (1997). How xDSL Supports Broadband Services to the Home. *IEEE Network 11*, 1 (Jan./Feb.), 14–23.

IEEE P802.11 VHT. (2010). IEEE 802.11 – VHT Study Group. http://www.ieee802.org/11/Reports/vht_update.htm.

IEEE P802.11ac VHT. (2010). IEEE 802.11 – Task Group AC. http://www.ieee802.org/11/Reports/tgac_update.htm.

IEEE P802.11ad VHT. (2010). IEEE 802.11 – Task Group AD. http://www.ieee802.org/11/Reports/tgad_update.htm.

IEEE P802.11e. (2005). Standard, Part 11: Wireless LAN Medium Access Control (MAC) and Physical Layer (PHY) Specifications: Medium Access Control (MAC) Quality of Service Enhancements.

IEEE P802.11n. (2009). Draft 9.0, Part 11: Wireless LAN Medium Access Control (MAC) and Physical Layer (PHY) Specifications: Enhancements for Higher Throughput.

IEEE P802.16m/D5. (2010). Draft 5.0, Amendment to IEEE Standard for Local and Metropolitan Area Networks.

Imashioya R., Syafei W. A., Nagao Y., *et al.* (2009). RTL Design of 1.2Gbps MIMO WLAN System and its Business Aspect. In *Proc., International Symposium on Communications and Information Technology (ISCIT)*. 296–301.

Ipakchi A., Albuyeh F. (2009). Grid of the Future. *IEEE Power & Energy Magazine 7*, 2 (March/April), 52–62.

Ishmael J., Bury S., Pezaros D., Race N. (2008). Deploying Rural Community Wireless Mesh Networks. *IEEE Internet Computing 12*, 4 (July/Aug.), 22–29.

ITU. (2009). The World in 2009: ICT Facts and Figures.

ITU-R M.1645. (2004). International Telecommunication Union-Recommendation (ITU-R).

ITU-R M.2134. (2008). Requirements Related to Technical System Performance for IMT-Advanced Radio Interface(s).

Jeon H. (2008). IEEE P802.11 Wireless LANs: 802.11 TGac Functional Requirements.

Jia Z., Yu J., Ellinas G., Chang G.-K. (2007). Key Enabling Technologies for Optical-Wireless Networks: Optical Millimeter-Wave Generation, Wavelength Reuse, and Architecture. *IEEE/OSA Journal of Lightwave Technology 25*, 11 (Nov.), 3452–3471.

Jian W., Chowdhury A., Jia Z., Estevez C. I., Chang G.-K. (2010). Energy-Efficient Multi-Access Technologies for Very-High-Throughput Avionic Millimeter Wave, Wireless Sensor Communication Networks. *IEEE/OSA Journal of Lightwave Technology 28*, 16 (Aug.), 2398–2405.

Jiang J., Handley M. R., Senior J. M. (2006). Dynamic Bandwidth Assignment MAC Protocol for Differentiated Services over GPON. *IET Electronics Letters 42*, 11 (May), 653–654.

Jin S., Choi M., Choi S. (2010). Performance Analysis of IEEE 802.16m Sleep Mode for Heterogeneous Traffic. *IEEE Communications Letters 14*, 5 (May), 405–407.

Jorguseski L., Fledderus E., Farserotu J., Prasad R. (2001). Radio Resource Allocation in Third Generation Mobile Communication Systems. *IEEE Communications Magazine 39*, 2 (Feb.), 117–123.

Kalantarisabet B., Mitchell J. E. (2006). MAC Constraints on the Distribution of 802.11 using Optical Fibre. In *Proc., European Conference on Wireless Technology*. 238–240.

Kani J.-I., Bourgart F., Cui A. *et al.* (2009). Next-Generation PON – Part I: Technology Roadmap and General Requirements. *IEEE Communications Magazine 47*, 11 (Nov.), 43–49.

Kanonakis K., Tomkos I., Pfeiffer T., Prat J., Kourtessis P. (2010). ACCORDANCE: A Novel OFDMA-PON Paradigm for Ultra-High Capacity Converged Wireline-Wireless Access Networks. In *Proc., International Conference on Transparent Optical Networks (ICTON)*. 1–4.

Kantarci B., Mouftah H. T. (2010). Reliable and Fast Restoration for a Survivable Wireless-Optical Broadband Access Network. In *Proc., International Conference on Transparent Optical Networks (ICTON)*. 1–4.

Katti S., Rahul H., Hu W., Katabi D., Medard M., Crowcroft J. (2008). XORs in the Air: Practical Wireless Network Coding. *IEEE/ACM Transactions on Networking 16*, 3 (June), 497–510.

Kautz W. R., Walker W. A. (2005). Drivers for Deep Fiber Access Solutions. *Tellabs White Paper*.

Kazovsky L. G., Shaw W.-T., Gutierrez D., Cheng N., Wong S.-W. (2007). Next-Generation Optical Access Networks. *IEEE/OSA Journal of Lightwave Technology, Special Issue on Convergence of Optical Wireless Access Networks 25*, 11 (Nov.), 3428–3442.

Kerpez K. J. (2002). DSL Spectrum Management Standard. *IEEE Communications Magazine 40*, 11 (Nov.), 116–123.

Kiese M., Georgieva E., Schupke D., Mukherjee B., Eberspächer J. (2009). Availability Evaluation in Hybrid Wireless-Optical Broadband Access Networks. In *Proc., IEEE International Conference on Communications (ICC)*. 1–6.

Kim B.-J., Shankaranarayanan N. K., Henry P. S., Schlosser K., Fong T. K. (1999). The AT&T Labs Broadband Fixed Wireless Field Experiment. *IEEE Communications Magazine 37*, 10 (Oct.), 56–62.

Kim G., Je Y., Kim S. (2009). An Adjustable Power Management for Optimal Power Saving in LTE Terminal Baseband Modem. *IEEE Transactions on Consumer Electronics 55*, 4 (Nov.), 1847–1853.

Kim M.-G., Shen G., Choi J. Y., *et al.* (2010). Distributed Antenna-Based EPON-WiMAX Integration and Its Cost-Efficient Cell Planning. *IEEE Journal on Selected Areas in Communications 28*, 6 (Aug.), 808–817.

Kim R. Y., Mohanty S. (2010). Advanced Power Management Techniques in Next-Generation Wireless Networks. *IEEE Communications Magazine 48*, 5 (May), 94–102.

Kishore S., Snyder L. V. (2010). Control Mechanisms for Residential Electricity Demand in SmartGrids. In *Proc., IEEE International Conference on Smart Grid Communications (SmartGridComm)*. 443–448.

Koonen T. (2006). Fiber to the Home/Fiber to the Premises: What, Where, and When. *Proceedings of the IEEE 94*, 5 (May), 911–934.

Kramer G. (2005). *Ethernet Passive Optical Networks*. McGraw-Hill, Inc.

Kramer G., Mukherjee B., Maislos A. (2003). Chapter 8: Ethernet Passive Optical Networks. In *IP over WDM: Building the Next Generation Optical Internet*, S. Dixit, ed. Wiley.

Kramer G., Mukherjee B., Pesavento G. (2001). Ethernet PON (EPON): Design and Analysis of an Optical Access Network. *Photonic Network Communications 3*, 3 (July), 307–319.

Kramer G., Mukherjee B., Pesavento G. (2002a). IPACT: A Dynamic Protocol for an Ethernet PON (EPON). *IEEE Communications Magazine 40*, 2 (Feb.), 74–80.

Kramer G., Mukherjee B., Ye Y., Dixit S., Hirth R. (2002b). Supporting Differentiated Classes of Service in Ethernet Passive Optical Networks. *OSA Journal of Optical Networking 1*, 8 (Aug.), 280–298.

Kuehl C. S. (1996). Applying Emerging Digital Video Interface Standards to Airborne Avionics Sensor and Digital Map Integrations: Benefits Outweigh the Initial Costs. In *Proc., SPIE*. 50–62.

Kuran M. S., Tugcu T. (2007). A Survey on Emerging Broadband Wireless Access Technologies. *Computer Networks 51*, 11 (Aug.), 3013–3046.

Kwan R., Leung C., Zhang J. (2009). Proportional Fair Multiuser Scheduling in LTE. *IEEE Signal Processing Letters 16*, 6 (June), 461–464.

Lamberth L. S., Laddu R. R., Harris D. *et al.* (2003). High-Resolution AM LCD Development for Avionic Applications. In *Proc., SPIE*. 70–79.

Lannoo B., Colle D., Pickavet M., Demeester P. (2007). Radio-over-Fiber-Based Solution to Provide Broadband Internet Access to Train Passengers. *IEEE Communications Magazine 45*, 2 (Feb.), 56–62.

Larmo A., Lindstrom M., Meyer M., *et al.* (2009). The LTE Link-Layer Design. *IEEE Communications Magazine 47*, 4 (Apr.), 52–59.

Le-Ngoc T., Leung V., Takats P., Garland P. (2003). Interactive Multimedia Satellite Access Communications. *IEEE Communications Magazine 41*, 7 (July), 78–85.

Lee J., Newaz S. H. S., Choi J. K., Lee G. M., Crespi N. (2009). QoS Mapping over Hybrid Optical and Wireless Access Networks. In *Proc., International Conference on Evolving Internet*. 139–141.

Lethien C., Loyez C., Vilcot J.-P. (2005). Potentials of Radio over Multimode Fiber Systems for the In-Buildings Coverage of Mobile and Wireless LAN Applications. *IEEE Photonics Technology Letters 17*, 12 (Dec.), 2793–2795.

Li B., Qin Y., Low C. P., Gwee C. L. (2007). A Survey on Mobile WiMAX. *IEEE Communications Magazine 45*, 12 (Dec.), 70–75.

Li Q., Li G., Lee W. *et al.* (2010a). MIMO Techniques in WiMAX and LTE: A Feature Overview. *IEEE Communications Magazine 48*, 5 (May), 86–92.

Li Q., Lin X., Zhang J., Roh W. (2009). Advancement of MIMO Technology in WiMAX: From IEEE 802.16d/e/j to 802.16m. *IEEE Communications Magazine 47*, 6 (June), 100–107.

Li X.-Y. (2009). Multicast Capacity of Wireless Ad Hoc Networks. *IEEE/ACM Transactions on Networking 17*, 3 (June), 950–961.

Li Y., Wang J., Qiao C., *et al.* (2010b). Integrated Fiber-Wireless (FiWi) Access Networks Supporting Inter-ONU Communications. *IEEE/OSA Journal of Lightwave Technology 28*, 5 (Mar.), 714–724.

Liao W.-K., Chen Y.-C. (2008). Supporting Vertical Handover between Universal Mobile Telecommunications System and Wireless LAN for Real-time Services. *IET Communications 2*, 1 (Jan.), 75–81.

Lim C., Nirmalathas A. (2010). Optical-Wireless Integration: Technologies for Physical Layer Networking. In *Proc., IEEE Wireless Communications & Networking Conference (WCNC), Workshop on Integrated Optical-Wireless Networks*. 1–6.

Lim C., Nirmalathas A., Bakaul M. *et al.* (2010). Fiber-Wireless Networks and Subsystem Technologies. *IEEE/OSA Journal of Lightwave Technology 28*, 4 (Feb.), 390–405.

Lin C.-T., Chen J., Peng P.-C. *et al.* (2007a). Hybrid Optical Access Network Integrating Fiber-to-the-Home and Radio-Over-Fiber Systems. *IEEE Photonics Technology Letters 19*, 8 (Apr.), 610–612.

Lin P., Wang T., Hu J., *et al.* (2007b). Optical-to-Wireless Integration Cost Modeling. In *Proc., IEEE Workhop on High Performance Switching and Routing (HPSR)*. 1–6.

Lin R. (2008). Next Generation PON in Emerging Networks. In *Proc., Optical Fiber Communication/National Fiber Optic Engineers Conference (OFC/NFOEC)*.

Lin W.-P., Kao M.-S., Chi S. (2003). A Reliable Architecture for Broad-Band Fiber-Wireless Access Networks. *IEEE Photonics Technology Letters 15*, 2 (Feb.), 344–346.

Lin Y., Wong V. (2006). Frame Aggregation and Optimal Frame Size Adaptation for IEEE 802.11n WLANs. In *Proc., IEEE Global Telecommunications Conference (GLOBECOM)*. 1–6.

Lin Y., Wong V. W. S. (2008). Utility-Optimal Cross-Layer Design for WLAN with MIMO Channels. In *Proc., IEEE International Conference on Communications (ICC)*. 2612–2617.

Lin Y.-J., Latchman H. A., Lee M., Katar S. (2002). A Power Line Communication Network Infrastructure for the Smart Home. *IEEE Wireless Communications 9*, 6 (Dec.), 104–111.

Liu G., Zhang J., Zhang P., *et al.* (2006). Evolution Map from TD-SCDMA to FuTURE B3G TDD. *IEEE Communications Magazine 44*, 3 (Mar.), 54–61.

Liu Y., Zhang S., Xu S., Zhang Y. (2009). Research on H.264/SVC Compressed Video Communication in 3G. In *Proc., 4th International Conference on Computer Science Education*. 327–332.

Lu K., Qian Y., Chen H.-H., Fu S. (2008). WiMAX Networks: From Access to Service Platform. *IEEE Network 22*, 3, 38–45.

Lun D. S., Médard M., Koetter R. (2006). Network Coding for Efficient Wireless Unicast. In *Proc., International Zurich Seminar on Communications*. 74–77.

Luo H., Ci S., Wu D., Wu J., Tang H. (2010). Quality-Driven Cross-Layer Optimized Video Delivery over LTE. *IEEE Communications Magazine 48*, 2 (Feb.), 102–109.

Luo Y., Wang T., Weinstein S. B., Cvijetic M., Nakamura S. (2006). Integrating Optical and Wireless Services in the Access Network. In *Proc., Optical Fiber Communication/National Fiber Optic Engineers Conference (OFC/NFOEC)*.

Luo Y., Yin S., Wang T. *et al.* (2007). QoS-Aware Scheduling over Hybrid Optical Wireless Networks. In *Proc., Optical Fiber Communication/National Fiber Optic Engineers Conference (OFC/NFOEC)*. 1–7.

Ma M., Zhu Y., Cheng T. H. (2003). A Bandwidth Guaranteed Polling MAC Protocol for Ethernet Passive Optical Networks. In *Proc., IEEE INFOCOM*. Vol. 1. 22–31.

Machuca C. M. (2006). Expenditures Study for Network Operators. In *Proc., International Conference on Transparent Optical Networks (ICTON)*. 18–22.

MacLeod J., Safavian S. R. (2008). FMC: Fixed-Mobile Convergence. *Bechtel Technology Journal 1*, 1 (Dec.), 57–75.

Maier M. (2009). WDM PONs and Beyond: The Road Ahead. *IEEE/OSA Journal of Optical Communications and Networking 1*, 4 (Sept.), C1–C16.

Maier M., Ghazisaidi N. (2010). QoS Provisioning Techniques for Future Fiber-Wireless (FiWi) Access Networks. *Future Internet, Special Issue on QoS in Wired and Wireless IP Networks 2*, 2 (Apr.), 126–155.

Maier M., Herzog M., Reisslein M. (2007). STARGATE: The Next Evolutionary Step Towards Unleashing the Potential of WDM EPONs. *IEEE Communications Magazine 45*, 5 (May), 50–56.

Majumder A., Caffery J. (2004). Power Line Communications: An Overview. *IEEE Potentials 23*, 8 (Oct./Nov.), 4–8.

Maxwell K. (1996). Asymmetric Digital Subscriber Line: Interim Technology for the Next Forty Years. *IEEE Communications Magazine 34*, 10 (Oct.), 100–106.

McGarry M. P., Maier M., Reisslein M. (2004). Ethernet PONs: A Survey of Dynamic Bandwidth Allocation (DBA) Algorithms. *IEEE Communications Magazine 42*, 8 (Aug.), S8–S15.

McGarry M. P., Maier M., Reisslein M. (2006). WDM Ethernet Passive Optical Networks. *IEEE Communications Magazine 44*, 2 (Feb.), S18–S25.

Miller K., Biermann T., Woesner H., Karl H. (2009). Network Coding in Passive Optical Networks. In *Proc., IEEE International Symposium on Network Coding*. 1–6.

Milosavljevic M., Kourtessis P., Senior J. M. (2010). Transparent Wireless Transmission over the ACCORDANCE Optical/Wireless Segment. In *Proc., International Symposium on Communication Systems Networks and Digital Signal Processing (CSNDSP)*. 138–142.

Molkdar D., Featherstone W., Larnbotharan S. (2002). An Overview of EGPRS: the Packet Data Component of EDGE. *Electronics & Communication Engineering Journal 14*, 1 (Feb.), 21–38.

Monath T., Kind M. (2007). Techno-Economics for Multi-Service Fixed Access Networks. *Netnomics 8*, 1–2 (Oct.), 91–103.

Mongha G., Pedersen K., Kovacs I., Mogensen P. (2008). QoS Oriented Time and Frequency Domain Packet Schedulers for The UTRAN Long Term Evolution. In *Proc., IEEE Vehicular Technology Conference (VTC)*. 2532–2536.

Moustafa M., Habib I., Naghshineh M., Guizani M. (2002). QoS-Enabled Broadband Mobile Access to Wireline Networks. *IEEE Communications Magazine 40*, 4 (Apr.), 50–56.

Mukai H., Hotta Y., Yokotani T., Takahashi A., Shimokasa K. (2009). PON with Automatic Protection Switching for High Reliable Communication. *Optical Switching and Networking 6*, 3 (July), 163–170.

Mukherjee B., Chowdhury P. (2009). Green Wireless-Optical Broadband Access Network (WOBAN). In *Proc., OSA Asia Communications and Photonics (ACP) Conference and Exhibition*. 1–2.

Muralidharan V., Wyglinski A. M., Wong W. (2007). HiFi-WiN: Hybrid Integrated Fiber-Wireless Networking for Broadband Metropolitan Area Access. In *Proc., Virginia Tech Symposium on Wireless Personal Communications*. 1–8.

Nadeem F., Kvicera V., Awan M. S., *et al.* (2009). Weather Effects on Hybrid FSO/RF Communication Link. *IEEE Journal on Selected Areas in Communications 27*, 9 (Dec.), 1687–1697.

Nandagopalan S., Cordeiro C., Fischer M. *et al.* (2009). MAC Channel Access in 60 GHz. https://mentor.ieee.org/802.11/dcn/09/11-09-0572-00-00ad-mac-channel-access-in-60-ghz.ppt.

Nava M. D., Del-Toso C. (2002). A Short Overview of the VDSL System Requirements. *IEEE Communications Magazine 40*, 12 (Dec.), 82–90.

Ni Q. (2005). Performance Analysis and Enhancements for IEEE 802.11e Wireless Networks. *IEEE Network 19*, 4 (Jul./Aug.), 21–27.

Nirmalathas A., Gamage P. A., Lim C., Novak D., Waterhouse R. (2010). Digitized Radio-Over-Fiber Technologies for Converged Optical Wireless Access Networks. *IEEE/OSA Journal of Lightwave Technology 28*, 16 (Aug.), 2366–2375.

Nirmalathas A., Gamage P. A., Lim C., *et al.* (2009). Digitized RF Transmission over Fiber. *IEEE Microwave Magazine 10*, 4 (June), 75–81.

Niyato D., Hossain E. (2007). Integration of WiMAX and WiFi: Optimal Pricing for Bandwidth Sharing. *IEEE Communications Magazine 45*, 5 (May), 140–146.

Niyato D., Hossain E., Kim D. I., Han Z. (2009). Relay-Centric Radio Resource Management and Network Planning in IEEE 802.16j Mobile Multihop Relay Networks. *IEEE Transactions on Wireless Communications 8*, 12 (Dec.), 6115–6125.

Ödling P., Magesacher T., Höst S., *et al.* (2009). The Fourth Generation Broadband Concept. *IEEE Communications Magazine 47*, 1 (Jan.), 63–69.

OECD. (2007). Broadband and the Economy. *Ministerial Background Report DSTI/ICCP/IE(2007)3/FINAL.*

OECD. (2009). Indicators of Broadband Coverage. *Working Party on Communication Infrastructures and Services Policy Report DSTI/ICCP/CISP(2009)3/FINAL.*

OECD. (2010). OECD Broadband Portal. http://www.oecd.org/sti/ict/broadband.

Papapanagiotou I., Toumpakaris D., Lee J., Devetsikiotis M. (2009). A Survey on Next Generation Mobile WiMAX Networks: Objectives, Features and Technical Challenges. *IEEE Communications Surveys & Tutorials.*

Papazian P. B., Hufford G. A., Achatz R. J., Hoffman R. (1997). Study of the Local Multipoint Distribution Service Radio Channel. *IEEE Transactions on Broadcasting 43*, 2, 175–184.

Pedersen K., Kolding T., Frederiksen F., *et al.* (2009). An Overview of Downlink Radio Resource Management for UTRAN Long-Term Evolution. *IEEE Communications Magazine 47*, 7 (July), 86–93.

Pelet E. R., Salt J. E., Wells G. (2004). Effect of Wind on Foliage Obstructed Line-of-Sight Channel at 2.5 GHz. *IEEE Transactions on Broadcasting 50*, 3 (Sept.), 224–232.

Peng M., Wang W., Chen H.-H. (2010). TD-SCDMA Evolution. *IEEE Vehicular Technology Magazine 5*, 2, 28–41.

Perahia E., Cordeiro C., Park M., Yang L. L. (2010). IEEE 802.11ad: Defining the Next Generation Multi-Gbps Wi-Fi. In *Proc., IEEE Consumer Communications and Networking Conference (CCNC).* 1–5.

Perkins C. E., Belding-Royer E. M. (1999). Ad-hoc On-Demand Distance Vector Routing. In *Proc., IEEE Workshop on Mobile Computing Systems and Applications (WMCSA).* 90–100.

Perkins C. E., Belding-Royer E. M., Das S. (2003). Ad Hoc On Demand Distance Vector (AODV) Routing. *IETF RFC 3561*.

Perkins C. E., Bhagwat P. (1994). Highly Dynamic Destination Sequenced Distance Vector Routing (DSDV) for Mobile Computers. In *Proc., SIGCOMM*. 234–244.

Peters S. W., Heath R. W. (2009). The Future of WiMAX: Multihop Relaying with IEEE 802.16j. *IEEE Communications Magazine 47*, 1 (Jan.), 104–111.

Peyravi H. (1999). Medium Access Control Protocols Performance in Satellite Communications. *IEEE Communications Magazine 37*, 3 (Mar.), 62–71.

Pleros N., Vyrsokinos K., Tsagkaris K., Tselikas N. D. (2009). A 60 GHz Radio-Over-Fiber Network Architecture for Seamless Communication With High Mobility. *IEEE/OSA Journal of Lightwave Technology 27*, 12 (June), 1957–1967.

Pospishny I., Vasyuk V., Romanchyk S., *et al.* (2010). 3GPP long term evolution (LTE). In *Proc., International Conference on Modern Problems of Radio Engineering, Telecommunications, and Computer Science (TCSET)*. 192.

Qian D., Cvijetic N., Hu J., Wang T. (2010). 108 Gb/s OFDMA-PON With Polarization Multiplexing and Direct Detection. *IEEE/OSA Journal of Lightwave Technology 28*, 4 (Feb.), 484–493.

Qiang Z., Hongbo Z. (2008). An Optimized AODV Protocol in Mobile Ad Hoc Network. In *Proc., International Conference on Wireless Communications, Networking and Mobile Computing (WiCOM)*. 1–4.

Qiu X. (2007). Powerful Talk. *IET Power Engineer 21*, 1, 38–43.

Racz A., Temesvary A., Reider N. (2007). Handover Performance in 3GPP Long Term Evolution (LTE) Systems. In *Proc., IST Mobile and Wireless Communications Summit*. 1–5.

Ray S. K., Pawlikowski K., Sirisena H. (2010). Handover in Mobile WiMAX Networks: The State of Art and Research Issues. *IEEE Communications Surveys & Tutorials 12*, 3, 376–399.

Reaz A., Ramamurthi V., Sarkar S. (2007). Capacity and Delay Aware Routing in Hybrid Wireless-Optical Broadband Access Network. In *Proc., International Symposium on Advanced Networks and Telecommunication Systems (ANTS)*. 1–2.

Reaz A., Ramamurthi V., Sarkar S. (2008a). Flow-Aware Channel Assignment in Multi-Radio Wireless-Optical Broadband Access Network. In *Proc., International Symposium on Advanced Networks and Telecommunication Systems (ANTS)*. 1–3.

Reaz A., Ramamurthi V., Sarkar S., *et al.* (2008b). CaDAR: an Efficient Routing Algorithm for Wireless-Optical Broadband Access Network. In *Proc., IEEE International Conference on Communications (ICC)*. 5191–5195.

Reaz A., Ramamurthi V., Sarkar S., *et al.* (2009). CaDAR: An Efficient Routing Algorithm for a Wireless-Optical Broadband Access Network (WOBAN). *IEEE/OSA Journal of Optical Communications and Networking 1*, 5 (Oct.), 392–403.

Reaz A., Ramamurthi V., Sarkar S., Ghosal D., Mukherjee B. (2008c). Hybrid Wireless-Optical Broadband Access Network (WOBAN): Capacity Enhancement for Wireless Access. In *Proc., IEEE GLOBECOM*. 1–5.

Rezgui J., Belbekkouche A., Hafid A. (2011). On Delay Performance and Burst Assembly for Wireless Mesh and Optical Burst Switching Converged Metro Area Network. *Mobile Networks and Applications 16*, 1 (Feb.), 122–133.

Richardson K. (2000). UMTS Overview. *Electronics & Communication Engineering Journal 12*, 3 (June), 93–100.

Ruth S. (2009). Green IT – More Than a Three Percent Solution? *IEEE Internet Computing 13*, 4 (July/Aug.), 74–78.

RVA Market Research & Consulting. (2009). Fiber-to-the-Home: North American Market Update. *Report for the FTTH Council.*

Samjani A. A. (2002). General Packet Radio Service (GPRS). *IEEE Potentials 21*, 2, 12–15.

Samukic A. (1998). UMTS Universal Mobile Telecommunications System: Development of Standards for the Third Generation. *IEEE Transactions on Vehicular Technology 47*, 4 (Nov.), 1099–1104.

Santacana E., Rackliffe G., Tang L., Feng X. (2010). Getting Smart. *IEEE Power & Energy Magazine 8*, 2 (March/April), 41–48.

Sarafi A. M., Tsiropoulos G. I., Cottis P. G. (2009). Hybrid Wireless-Broadband over Power Lines: A Promising Broadband Solution in Rural Areas. *IEEE Communications Magazine 47*, 11 (Nov.), 140–147.

Sarikaya B. (2000). Packet Mode in Wireless Networks: Overview of Transition to Third Generation. *IEEE Communications Magazine 38*, 9 (Sept.), 164–172.

Sarkar S., Dixit S., Mukherjee B. (2007a). Hybrid Wireless-Optical Broadband-Access Network (WOBAN): A Review of Relevant Challenges. *IEEE/OSA Journal of Lightwave Technology 25*, 11 (Nov.), 3329–3340.

Sarkar S., Mukherjee B., Dixit S. (2006). Optimum Placement of Multiple Optical Network Units (ONUs) in Optical-Wireless Hybrid Access Networks. In *Proc., Optical Fiber Communication/National Fiber Optic Engineers Conference (OFC/NFOEC).* 1–3.

Sarkar S., Yen H.-H., Dixit S., Mukherjee B. (2007b). RADAR: Risk-and-Delay Aware Routing Algorithm in a Hybrid Wireless-Optical Broadband Access Network (WOBAN). In *Proc., Optical Fiber Communication/National Fiber Optic Engineers Conference (OFC/NFOEC).* 1–3.

Sarkar S., Yen H.-H., Dixit S., Mukherjee B. (2008). A Novel Delay-Aware Routing Algorithm (DARA) for a Hybrid Wireless-Optical Broadband Access Network (WOBAN). *IEEE Network 22*, 3 (Jan./Feb.), 20–28.

Sarkar S., Yen H.-H., Dixit S., Mukherjee B. (2009). Hybrid Wireless-Optical Broadband Access Network (WOBAN): Network Planning Using Lagrangean Relaxation. *IEEE/ACM Transactions on Networking 17*, 4 (Aug.), 1094–1105.

Schiller J. (2003). *Mobile Communications*, Second edn. Addison-Wesley.

Schütz G., Correia N. S. C. (2009). A Heuristic for Fault-Tolerance Provisioning in Multi-Radio Hybrid Wireless-Optical Broadband Access Network. In *Proc., International Conference on Transparent Optical Networks (ICTON).* 1–4.

Shao X., Yeo Y. K., Ngoh L. H., *et al.* (2010). Availability-Aware Routing for Large-Scale Hybrid Wireless-Optical Broadband Access Network. In *Proc., Optical Fiber Communication/National Fiber Optic Engineers Conference (OFC/NFOEC).* 1–3.

Shaw W.-T., Kalogerakis G., Wong S.-W. *et al.* (2006). MARIN: Metro-Access Ring Integrated Network. In *Proc., IEEE GLOBECOM.* 1–5.

Shaw W.-T., Wong S.-W., Cheng N. *et al.* (2008). Reconfigurable Optical Backhaul and Integrated Routing Algorithm for Load Balancing in Hybrid Optical-Wireless Access Networks. In *Proc., IEEE International Conference on Communications (ICC).* 5697–5701.

Shaw W.-T., Wong S.-W., Cheng N. *et al.* (2007a). Hybrid Architecture and Integrated Routing in a Scalable Optical-Wireless Access Network. *IEEE/OSA Journal of Lightwave Technology, Special Issue on Convergence of Optical Wireless Access Networks 25*, 11 (Nov.), 3443–3451.

Shaw W.-T., Wong S.-W., Cheng N., Kazovsky L. G. (2007b). MARIN Hybrid Optical-Wireless Access Network. In *Proc., Optical Fiber Communication/National Fiber Optic Engineers Conference (OFC/NFOEC)*. 1–3.

Shaw W.-T., Wong S.-W., Yen S.-H., Kazovsky L. G. (2009). An Ultra-Scalable Broadband Architecture for Municipal Hybrid Wireless Access Using Optical Grid Network. In *Proc., Optical Fiber Communication/National Fiber Optic Engineers Conference (OFC/NFOEC)*. 1–3.

Shayani D., Machuca C. M., Jäger M., Gladisch A. (2008). Cost Analysis of the Service Migration Problem between Communication Platforms. In *Proc., IEEE Network Operations and Management Symposium (NOMS)*. 734–737.

Shea D. P., Mitchell J. E. (2007). Long-Reach Optical Access Technologies. *IEEE Network 21*, 5 (Sept./Oct.), 5–11.

Shen G., Tucker R. S., Chae C.-J. (2007). Fixed Mobile Convergence Architectures for Broadband Access: Integration of EPON and WiMAX. *IEEE Communications Magazine 45*, 8 (Aug.), 44–50.

Sherif S. R., Hadjiantonis A., Ellinas G., Assi C., Ali M. A. (2004). A Novel Decentralized Ethernet-Based PON Access Architecture for Provisioning Differentiated QoS. *IEEE/OSA Journal of Lightwave Technology 22*, 11 (Nov.), 2483–2497.

Shumate P. W. (2008). Fiber-to-the-Home: 1977–2007. *IEEE/OSA Journal of Lightwave Technology 26*, 9 (May), 1093–1103.

Sim S., Han S.-J., Park J.-S., Lee S.-C. (2009). Seamless IP Mobility Support for Flat Architecture Mobile WiMAX Networks. *IEEE Communications Magazine 47*, 6 (June), 142–148.

Skordoulis D., Ni Q., Chen H., *et al.* (2008). IEEE 802.11n MAC Frame Aggregation Mechanisms for Next-Generation High-Throughput WLANs. *IEEE Wireless Communications 15*, 1 (Feb.), 40–47.

So-In C., Jain R., Tamimi A.-K. (2009). Scheduling in IEEE 802.16e Mobile WiMAX Networks: Key Issues and a Survey. *IEEE Journal on Selected Areas in Communications 27*, 2 (Feb.), 156–171.

Somemura Y. (2010). Key Roles of Green Technology for Access Network Systems. In *Proc., OSA Optical Fiber Communication/National Fiber Optic Engineers Conference (OFC/NFOEC)*, paper OTuO6.

Song K. B., Chung S. T., Ginis G., Cioffi J. M. (2002). Dynamic Spectrum Management for Next-Generation DSL Systems. *IEEE Communications Magazine 40*, 10 (Oct.), 101–109.

Sood V. K., Fischer D., Eklund J. M., Brown T. (2009). Developing a Communication Infrastructure for the Smart Grid. In *Proc., IEEE Electrical Power & Energy Conference*. 1–7.

Stotts L. B., Andrews L. C., Cherry P. C. *et al.* (2009). Hybrid Optical RF Airborne Communications. *Proceedings of the IEEE 97*, 6 (June), 1109–1127.

Subramanian A. P., Buddhikot M. M., Miller S. (2006). Interference Aware Routing in Multi-Radio Wireless Mesh Networks. In *Proc., IEEE Workshop on Wireless Mesh Networks (WiMesh)*. 55–63.

Syafei W. A., Nagao Y., Kurosaki M., Sai B., Ochi H. (2009). Design of 1.2 Gbps MIMO WLAN System for 4K Digital Cinema Transmission. In *Proc., IEEE International Symposium on Personal, Indoor and Mobile Radio Communications*. 207–211.

Talli G., Townsend P. D. (2006). Hybrid DWDM–TDM Long-Reach PON for Next-Generation Optical Access. *IEEE/OSA Journal of Lightwave Technology 24*, 7 (July), 2827–2834.

Tanaka K., Agata A., Horiuchi Y. (2010). IEEE 802.3av 10G-EPON Standardization and Its Research and Development Status. *IEEE/OSA Journal of Lightwave Technology 28*, 4 (Feb.), 651–661.

Tang P. K., Ong L. C., Alphones A., Luo B., Fujise M. (2004). PER and EVM Measurements of a Radio-Over-Fiber Network for Cellular and WLAN System Applications. *IEEE/OSA Journal of Lightwave Technology 22*, 11 (Nov.), 2370–2376.

Tran A. V., Chae C. J., Tucker R. S. (2005). Ethernet PON or WDM PON: A Comparison of Cost and Reliability. In *Proc., IEEE TENCON*. 1–6.

Tucker R. S., Parthiban R., Baliga J., *et al.* (2009). Evolution of WDM Optical IP Networks: A Cost and Energy Perspective. *IEEE/OSA Journal of Lightwave Technology 27*, 3 (Feb.), 243–252.

Vaughan-Nichols S. J. (2008). Mobile WiMax: The Next Wireless Battleground. *Computer 41*, 6 (June), 16–18.

Verbrugge S., Pasqualini S., Westphal F.-J. *et al.* (2005). Modeling Operational Expenditures for Telecom Operators. In *Proc., Optical Network Design and Modelling (ONDM)*. 455–466.

Vermesan I., Moldovan A., Palade T., Colda R. (2010). Multi Antenna STBC Transmission Technique Evaluation under IEEE 802.11n Conditions. In *Proc., International Conference on Microwave Techniques (COMITE)*. 51–54.

Wagner R. E., Igel J. R., Whitman R., *et al.* (2006). Fiber-Based Broadband-Access Deployment in the United States. *IEEE/OSA Journal of Lightwave Technology 24*, 12 (Dec.), 4526–4540.

Wang F., Ghosh A., Sankaran C., *et al.* (2008). Mobile WiMAX Systems: Performance and Evolution. *IEEE Communications Magazine 46*, 10 (Oct.), 41–49.

Wang J., Wu K., Li S., Qiao C. (2010). Performance Modeling and Analysis of Multi-Path Routing in Integrated Fiber-Wireless Networks. In *Proc., Mini-Conference at IEEE INFOCOM*. 1–5.

Weldon M. K., Zane F. (2003). The Economics of Fiber to the Home Revisited. *Bell Lab Technical Journal 8*, 1 (July), 181–206.

Wells J. (2009). Faster Than Fiber: The Future of Multi-Gb/s Wireless. *IEEE Microwave Magazine 10*, 3 (May), 104–112.

Wittig M. (2000). Satellite Onboard Processing for Multimedia Applications. *IEEE Communications Magazine 38*, 6 (June), 134–140.

Wong S.-W., Campelo D. R., Cheng N. *et al.* (2009). Grid Reconfigurable Optical-Wireless Architecture for Large Scale Municipal Mesh Access Network. In *Proc., IEEE GLOBECOM*. 1–6.

Wong S.-W., Shaw W.-T., Balasubramaniam K., Cheng N., Kazovsky L. G. (2007). MARIN: Demonstration of a Flexible and Dynamic Metro-Access Integrated Architecture. In *Proc., IEEE GLOBECOM*. 2173–2177.

Wu J., Wu J.-S., Tsao H.-W. (1994). A Fiber Distribution System for Microcellular Radio. *IEEE Photonics Technology Letters 6*, 9 (Sept.), 1150–1152.

Wu M., Makharia S., Liu H., Li D., Mathur S. (2009). IPTV Multicast Over Wireless LAN Using Merged Hybrid ARQ With Staggered Adaptive FEC. *IEEE Transactions on Broadcasting 55*, 2, 363–374,.

Xia L., Liu Z., Chang Y., Sun P. (2009). An Improved AODV Routing Protocol Based on the Congestion Control and Routing Repair Mechanism. In *Proc., International Conference on Communications and Mobile Computing (CMC)*. 259–262.

Xu K., Sun X., Yin J. *et al.* (2010). Enabling ROF Technologies and Integration Architectures for In-Building Optical-Wireless Access Networks. *IEEE Photonics Journal 2*, 2 (Apr.), 102–112.

Yan Y., Yu H., Wessing H., Dittmann L. (2009). Enhanced Signaling Scheme with Admission Control in the Hybrid Optical Wireless (HOW) Networks. In *Proc., IEEE INFOCOM Workshops*. 1–6.

Yang S.-R., Kao C.-C., Kan W.-C., Shih T.-C. (2010). Handoff Minimization Through a Relay Station Grouping Algorithm With Efficient Radio-Resource Scheduling Policies for IEEE 802.16j Multihop Relay Networks. *IEEE Transactions on Vehicular Technology 59*, 5 (June), 2185–2197.

Yang Y., Wang J., Kravets R. (2005). Designing Routing Metrics for Mesh Networks. In *Proc., IEEE Workshop on Wireless Mesh Networks (WiMesh)*. 1–9.

Yoneda S. (2010). Design of Low-Carbon Electric and Communications Infrastructure. In *Proc., IEEE International Conference on Smart Grid Communications (SmartGridComm)*. 472–476.

Yu J., Chang G.-K., Jia Z. *et al.* (2010). Cost-Effective Optical Millimeter Technologies and Field Demonstrations for Very High Throughput Wireless-Over-Fiber Access Systems. *IEEE/OSA Journal of Lightwave Technology 28*, 16 (Aug.), 2376–2397.

Yubin Z., Li H., Ruitao X., Yaojun Q., Yuefeng J. (2009). Wireless Protection Switching for Video Service in Wireless-Optical Broadband Access Network. In *Proc., IEEE International Conference on Broadband Network & Multimedia Technology (IC-BNMT)*. 760–764.

Zhang J., Ansari N., Luo Y., Effenberger F., Ye F. (2009). Next-Generation PONs: A Performance Investigation of Candidate Architectures for Next-Generation Access Stage 1. *IEEE Communications Magazine 47*, 8 (Aug.), 49–57.

Zhang L., An E.-S., Youn C.-H., Yeo H.-G., Yang S. (2003). Dual DEB-GPS Scheduler for Delay-Constraint Applications in Ethernet Passive Optical Networks. *IEICE Transactions on Communications E86-B*, 5 (May), 1575–1584.

Zheng J., Mouftah H. T. (2005). Media Access Control for Ethernet Passive Optical Networks: An Overview. *IEEE Communications Magazine 43*, 2 (Feb.), 145–150.

Zheng K., Fan B., Ma Z., Liu G., Shen X., Wang W. (2009). Multihop Cellular Networks toward LTE-Advanced. *IEEE Vehicular Technology Magazine 4*, 3, 40–47.

Zheng N., Wigard J. (2008). On the Performance of Integrator Handover Algorithm in LTE Networks. In *Proc., IEEE Vehicular Technology Conference (VTC)*. 1–5.

Zheng Z., Wang J., Wang J. (2009a). A Study of Network Throughput Gain in Optical-Wireless (FiWi) Networks Subject to Peer-to-Peer Communications. In *Proc., IEEE International Conference on Communications (ICC)*. 1–6.

Zheng Z., Wang J., Wang X. (2009b). ONU Placement in Fiber-Wireless (FiWi) Networks Considering Peer-to-Peer Communications. In *Proc., IEEE GLOBECOM*. 1–7.

Index

Printed in the United States
By Bookmasters